Pozzolanic and
Cementitious Materials

Advances in Concrete Technology

A series edited by V. M. Malhotra
Advanced Concrete Technology Program, CANMET
Ottawa, Ontario, Canada

This series will consist of approximately ten short, sharply focused tracts, each one covering one of the many aspects of concrete technology: materials, construction, and testing. The goal of this series is to provide a convenient, practical, and current source on concrete technology for practicing civil and structural engineers, concrete technologists, manufacturers suppliers, and contractors involved in construction and maintenance of concrete structures.

Volume 1
Pozzolanic and Cementitious Materials
V. M. Malhotra and P. Kumar Mehta

In Preparation

Fly Ash in Cement and Concrete
R. C. Joshi and R. P. Lohtia

Fiber Reinforced Concrete
Colin D. Johnston

Alkali-Aggregate Reaction in Concrete
Marc-André Bérubé and Benoit Fournier

This book is part of a series. The publisher will accept continuation orders which may be cancelled at any time and which provide for automatic billing and shipping of each title in the series upon publication. Please write for details.

Pozzolanic and Cementitious Materials

V. M. Malhotra
Advanced Concrete Technology, CANMET
Ottawa, Ontario
Canada

and

P. Kumar Mehta
University of California at Berkeley
USA

GORDON AND BREACH PUBLISHERS

Australia Canada China France Germany India Japan
Luxembourg Malaysia The Netherlands Russia Singapore
Switzerland Thailand United Kingdom

Emmaplein 5
1075 AW Amsterdam
The Netherlands

British Library Cataloguing in Publication Data

Malhotra, V. M.
  Pozzolanic and Cementitious Materials. –
  (Advances in Concrete Technology, ISSN
  1024–5308;Vol.1)
  I. Title II. Mehta, P. K. III. Series
  624.1833

  ISBN   2–88449–235–6   (hardcover)
         2–88449–211–9   (softcover)

CONTENTS

Contents

PREFACE

Portland-cement concrete is the most widely used construction material today. From massive dams to elegant reinforced and prestressed buildings, concrete finds application in a variety of structures. Compared to metals, ceramics, and polymeric materials, concrete is always less expensive, possesses adequate strength and durability, and requires less energy to produce. The energy associated with the production of a typical concrete mixture can be reduced even further. Portland cement is the most energy-intensive component of a concrete mixture, whereas pozzolanic and cementitious by-products from thermal power production and metallurgical operations require little or no expenditure of energy. Therefore, as a cement substitute, typically from 20% to 60% cement replacement by mass, the use of such by-products in the cement and concrete industry can result in substantial energy savings.

Concrete mixtures containing pozzolanic and cementitious materials exhibit superior durability to thermal cracking and aggressive chemicals This explains the increasing worldwide trend toward utilization of pozzolanic and cementitious materials either in the form of blended portland cements or as direct additions to portland-cement concrete during the mixing operation.

The authors believe that in the future, further increases in the utilization of pozzolanic and cementitious materials in concrete will come from the realization of ecological benefits associated with such usage. The total amount of pozzolanic and cementitious by-products generated by the industry worldwide every year exceeds 500 million tonnes. Many of the by-products contain toxic elements which can be hazardous if not disposed of in a safe manner. These toxic elements find their way into groundwater when industrial wastes are dumped or used as landfill or roadbase. The cement and concrete industry provides a preferred home for the by-product pozzolanic and cementitious materials because most of the toxic metals present in the by-product can be permanently bound into the portland cement hydration products.

The pozzolanic and cementitious materials discussed in this book include not only industrial by-products such as fly ash, blast-furnace slag, silica fume, and rice-husk ash, but also materials from natural sources. The terms "mineral admixtures" or "supplementary cementing materials" are often used to cover all pozzolanic and cementitious materials, irrespective of their source. Definitions of the materials covered in this book, history of their development, and principal benefits derived from their use are included in the first chapter. Main sources of production as well as current rates of utilization in major countries of the world are discussed in the second chapter. In the third chapter, the physical and chemical properties, including the mineralogical composition and particle characteristics, are given. The mechanisms by which mineral admixtures improve the properties of concrete are covered in chapter 4. It is hoped that the information in this chapter will be helpful to practicing engineers in understanding why the potential benefits expected from the use of pozzolanic and cementitious materials are not always realized.

The next three chapters contain selected laboratory and field data on the performance of concrete containing each mineral admixture type. The effect of mineral admixtures on properties of concrete and hardened concrete are discussed in chapters 5 and 6, respectively. Durability aspects of concrete containing mineral admixtures are covered in chapter 7. A brief discussion on the selection of materials and mixture proportions is presented in chapter 8. Test methods and standard specifications, including their significance, are given in chapter 9. The U.S., Canadian, and British standards are highlighted because many countries of the world follow these standards or use similar standards. This concluding chapter contains a proposal for a universal standard covering all mineral admixtures.

This book is meant primarily for use by practicing engineers; therefore, it is not intended to be a comprehensive or scholarly treatise on the subject. An attempt is made here to present only the essential information in a simple and straightforward manner. The authors will have achieved their limited objective if construction engineers and concrete technologists —usually too busy to go through research papers and conference proceedings—find this book useful and start incorporating more pozzolanic and cementitious materials into concrete mixtures.

CHAPTER 1

Introduction

The use of pozzolanic and cementitious materials in the cement and concrete industries has risen sharply during the last fifty years. For a variety of reasons, discussed later, the potential for future use is even greater. It is predicted that in the near future a concrete mixture without pozzolanic or cementitious materials would be an exception rather than the rule. Therefore, it is important for practicing civil engineers to acquire a basic knowledge of the types of materials available, their characteristics, and their influence on properties of concrete.

Definitions

Pozzolanic and cementitious admixtures are generally classified under the term, **mineral admixtures**. Strictly speaking, the term **admixture** refers to any material other than water, aggregates and cement, used as a concrete ingredient, and added to the batch immediately before or during mixing. When a pozzolanic or a cementitious material is interground or blended during the manufacture of cement, as in the case of ASTM Type IP and Type IS cements (portland pozzolan and portland blast-furnace slag cements, respectively), it is called a **mineral addition**. As the advantages associated with the use of a pozzolanic or cementitious material remain essentially the same, irrespective of the method by which it has been incorporated into a concrete mixture, the term mineral admixture is used in a broad sense in this book.

1

A **pozzolan** is a *siliceous or siliceous and aluminous material, which in itself possesses little or no cementitious property but which will, in finely divided form and in the presence of moisture, chemically react with calcium hydroxide at ordinary temperature to form compounds possessing cementing properties.* Volcanic ashes, calcined clays, and pulverized coal ash from thermal power plants are among the commonly used pozzolanic materials.

To develop cementing action, it is evident that a pozzolan has to be mixed either with lime or with portland cement. (Note that calcium hydroxide is one of the products of portland cement-water interaction.) However, *there are some finely divided and non-crystalline or poorly crystalline materials similar to pozzolans but containing sufficient calcium to form compounds which possess cementing properties after interaction with water. These materials are classified as* **cementitious.** Examples of cementitious mineral admixtures are granulated blast-furnace slag and high-calcium fly ash. **Granulated blast-furnace slag** is a nonmetallic product, consisting essentially of silicates and aluminosilicates of calcium and magnesium. When the molten slag is quenched rapidly with a large quantity of water, it forms a granular material, which is essentially glassy or noncrystalline.

HISTORY OF DEVELOPMENT

According to Lea (1), lime-pozzolan composites were used as a cementing material for construction of structures throughout the Roman empire. One source of pozzolan was an ash produced by the volcanic eruption of Mount Vesuvius in 79 A.D. which destroyed Pompei, Herculanum and several other towns along the Bay of Naples. In fact, it was in Italy that the term, "pozzolan" was first used to describe the volcanic ash mined at Pozzuoli, a village near Naples. It should be noted that a similar natural ash, formed as a result of volcanic eruption in Santorini Island around 1500 B.C., had been in use for making lime-pozzolan mortars in Greece much before the word pozzolan was coined.

Another pozzolanic material, Rheinisch trass, which is a volcanic tuff, was used extensively in Germany during the Roman period. There is also evidence that natural materials, such as volcanic ash and tuffs, were not the sole sources of pozzolans in lime-pozzolan composites used for the construction of ancient structures. According to Lea (1), some Minoan structures of Crete Island, built around 1500–2000 B.C., contained crushed and ground potsherds (calcined clay products) in a lime mortar. In fact, not only Greek and Roman but also Indian and Egyptian civilizations were familiar with the water-resisting property of mortars and concretes made with lime-pozzolanic cements, the source of pozzolan being calcined clay from crushed bricks, tile, and pottery.

The discovery and use of hydraulic limes (impure limes containing substantial amounts of calcined clay) during the 18th century was a forerunner of the invention of portland cement in 1824. Due to faster setting and hardening characteristics of portland cement, it quickly became the favorite cementing material of the construction industry instead of lime-pozzolan cements. However, due to technological, economical, and ecological considerations discussed below, large quantities of pozzolanic materials continue to be in use today in the form of mineral admixtures for the portland cement and concrete industries. Another point which should be mentioned here is that industrial by-products, such as fly ash from thermal power plants, and slag as well as silica fume from metallurgical operations, are being increasingly substituted for natural pozzolans and calcined clays.

Granulated blast-furnace slag was first developed in Germany in 1853. In most countries of the world, it is incorporated as an ingredient of blended portland-blast furnace slag cements containing 25 to 70% slag by mass. In the U.S., where substantial quantities of the granulated material were not available until recently, most of the granulated blast-furnace slag is now being marketed as an admixture for concrete. It has been shown that high-calcium fly ash produced by the combustion of subbituminous and lignite coals is mineralogically similar to granulated blast-furnace slag (2). A predominant constituent of both is a calcium-rich, alumino-silicate glass which contributes to their self-cementing property.

Silica fume is a relatively new material, which has come into increasing use in the concrete industry since the 1980s. Due to its extremely fine particle size and amorphous nature, it is a highly reactive pozzolan. Two other products, which are highly reactive pozzolans, are still under development. One is rice-husk ash (called rice-hull ash in the U.S.) produced by controlled combustion of rice husks. Rice husks represent one of the largest disposable crop residues in the world. When used as a fuel in a thermal power plant, the combustion of rice husks can yield an amorphous silica ash of very high surface area. The second product, called metakaolin, is made by low-temperature calcination of high-purity kaolin clay. The calcined product contains silica and alumina in an amorphous form, and is ground to very fine particle size.

BENEFITS FROM THE USE OF MINERAL ADMIXTURES

The benefits derived from the use of mineral admixtures in the cement and concrete industries can be divided into three categories: functional or engineering benefits, economic benefits, and ecological benefits.

Engineering Benefits: First, the incorporation of finely divided particles into a concrete mixture tends to improve the workability, and reduce the water requirement at a given consistency (except for materials with a very high surface area, such as silica fume). Secondly, in general, there is an enhancement of ultimate strength, impermeability, and durability to chemical attack. Thirdly, an improved resistance to thermal cracking is obtained due to the lower heat of hydration of blended cements and increased tensile strain capacity of concrete containing mineral admixtures. The mechanisms on how mineral admixture improve the properties of concrete are discussed in Chapter 4. Detailed data showing the effects of mineral admixtures on the properties of concrete are presented in Chapters 5, 6 and 7.

Economic Benefits: Typically, portland cement represents the most expensive component of a concrete mixture. As it is a highly energy-intensive material, the increasing energy costs in recent years have been reflected in correspondingly higher cement costs. On the other hand, most of the pozzolanic and cementitious materials in use today are industrial by-products, which require relatively little or no expenditure of energy for use as mineral admixtures. Obviously, when used as a partial cement replacement, typically in the range of 20 to 60% cement by mass, mineral admixtures can result in substantial energy and cost savings.

Ecological Benefits: The total volume of pozzolanic and cementitious by-products generated every year by thermal power plants and metallurgical furnaces exceeds 500 million tons. Many of these by-products contain toxic elements which can be hazardous to human health if not disposed in a safe manner. Thus dumping into lakes, streams or landfills and use as a roadbase is not a safe practice because the toxic elements can find their way into groundwater. The cement and concrete industries provide a preferred vehicle for disposal of by-product mineral admixtures because most of the harmful metals can be safely incorporated into the hydration products of cement.

Furthermore, every ton of portland cement production is accompanied by a similar amount of carbon dioxide as a by-product, which is released into the environment. This means that today's portland cement production of 1 billion tons/year is already responsible for substantial environmental loading by CO_2, which is a primary factor in the "greenhouse" effect. In the interest of environmental protection, it is therefore desirable that the rising cement demand in the world is met by higher rates of utilization of mineral admixtures used as supplementary cementing materials rather than by further increases in the production of portland cement, which was the practice in the past.

CHAPTER 2

Production, Sources, and Utilization

This chapter covers the wide ranging sources from which mineral admixtures are derived. An attempt is also made to give the current production and utilization rates in various parts of the world. However, due to rapid changes in industrial and economic conditions, the production and utilization rates of individual mineral admixtures are subject to wide fluctuations. For example, traditional mineral admixtures were derived mostly from natural volcanic materials and calcined clays or shales. With the availability of industrial by-products, such as coal ashes and metallurgical slags, in many countries the economic and ecological considerations have forced the substitution of natural materials by industrial by-products.

MINERAL ADMIXTURES FROM NATURAL SOURCES

Except for diatomaceous earth, all natural pozzolanic materials are derived from volcanic rocks. Volcanic eruptions hurl into the atmosphere large quantities of molten lava, which is composed mainly of aluminosilicates. Quick cooling of lava results in the formation of vitreous phases (glass) with disordered structure or poorly crystalline minerals. Also, the escaping gases and water vapor impart to the volcanic material a porous texture with a high surface area. A combination of glassy or poorly crystalline structure and high surface area is

7

the cause for the reactivity of aluminosilicate phases present in volcanic ash with calcium hydroxide at normal temperature. Santorin Earth of Greece, Bacoli Pozzolan of Italy, and Shirasu Pozzolan of Japan are examples of unaltered **volcanic glass**. Santorin Earth was used as a mineral admixture for the concrete lining of Suez Canal. It is reported that portland cements made in Greece routinely contain 10% Santorin Earth.

Hydrothermal alterations convert volcanic glass to **zeolitic trass or tuff**, which has a chemical composition of the type, $(Na_2Ca)O \cdot Al_2O_3 . 4SiO_2 . xH_2O$. Zeolitic minerals in finely ground condition show pozzolanic behavior because they react with lime by a base-exchange process to form cementitious products. Zeolitic tuffs and trass are found in many countries including Germany, Italy, China, and Russia. Although only a small amount of trass (approx. 50,000 tons/year) is being used for the production of portland-pozzolan cements in Germany, the quantities used in Italy and China are much larger. According to recent reports, Italy consumes yearly about 3 million tons, and China about 5 million tons of zeolitic tuffs as a component of blended portland cements.

Clay minerals, which are not pozzolanic, are formed by progressive alteration of volcanic glass. The crystalline structure of aluminosilicate minerals present in a clay or shale can be destroyed by heating to 700–800°C; the resulting product, i.e. **calcined clays and shales** are pozzolanic. In fact, before the advent of fly ash, calcined clays and shales have been frequently employed as a mineral admixture for construction of many mass concrete structures. **Metakaolin**, a recent product on the market is made from a high-purity kaolin clay by low-temperature calcination. The product is pulverized to very fine particle size (average 1–2 μm) to make it highly pozzolanic. Currently, small amounts of this material are being produced near Atlanta in the U.S. According to Caldarone et al. (3), a kaolinitic clay is washed to remove unreactive impurities and the resulting high-purity kaolin is calcined at a specific temperature to produce a highly reactive metakaolin. The product, ground to an average particle size 1.5 µm, is white in color and, therefore more suitable for use in architectural concrete. Ambroise et al. (4) from France have reported the properties of metakaolin produced in a small rotary kiln by calcination at 700–800°C of a clay containing 95% kaolinite.

Diatomaceous Earth consists of amorphous hydrated silica derived from the skeleton of diatoms, which are tiny water plants with cell walls composed of silica shells. The high-purity material is pozzolanic, but it is usually found heavily contaminated with clays and therefore must be calcined to enhance pozzolanic reactivity. Large deposits of diatomaceous earth are found in several parts of the world, including Algeria, California, Denmark, and France. In the past, the material found only limited applications in the cement and concrete industries because of the large water requirement to obtain proper consistency.

COAL ASHES

The combustible constituents in coal are composed mainly of carbon, hydrogen and oxygen; nitrogen and sulfur being the minor elements. In addition, depending upon the grade of coal, substantial amounts of non-combustible impurities, from about 10 to 40 percent, are usually present in the form of clay, shale, quartz, feldspar, and limestone. Due to superior burning efficiency most thermal power plants use coal-fired boilers which consume coal ground to a fineness of more than 75% particles passing the No. 200–mesh sieve (74 µm). As the fuel travels through the high-temperature zone in the furnace, the volatile matter and carbon are burnt off, whereas most of the mineral impurities are fused and remain suspended in the flue gas. Upon leaving the combustion zone the molten ash particles are cooled rapidly (e.g. from 1500°C to 200°C in a few seconds) and they solidify as spherical, glassy particles. Some of the fused matter agglomerates to form **bottom ash**, but most of it flies out with the flue gas stream and is therefore called, **fly ash**. Subsequently, the fly ash is removed from the flue gas by a series of mechanical separators and electrostatic precipitators or bag filters. Typically, the ratio of fly ash to bottom ash is: 70:30 in wet-bottom boilers, or 85:15 in dry-bottom boilers. Due to its unique mineralogical and granulometric characteristics, fly ash generally does not need any processing before use as a mineral admixture. Bottom ash is much coarser, less reactive, and therefore requires fine grinding to develop a pozzolanic property.

In many countries, efforts are being made to reduce substantially the NO_x emissions from furnaces using fossil fuel. For instance, in the Netherlands before the end of 1995, about 70% of the fly ash is expected to originate from plants using NO_x-emission-reducing techniques. According to a report from the Dutch Center for Civil Engineering Research and Codes (5), an important contribution to the NO_x reduction is made by lowering the peak burning temperature. The temperature is lowered to such a level that ash, containing minerals with high melting point, does not always fuse completely and, therefore, fewer round-shaped particles and more conglomerates of particles sintered together are formed. As the amount of fly ash particles smaller than 10 μm is substantially reduced, the reactivity and the particle-packing effect (as discussed in Chapter 4) are adversely affected. This type of fly ash is, therefore, of somewhat lower quality than the fly ash from conventional furnaces (5).

The U.S., China, India, Russia, South Africa, and Untied Kingdom are among the largest fly ash producing countries of the world. The annual fly ash production in the world is estimated at about 450 million tons; the current (1993) U.S. production, according to the American Coal Ash Association, is about 48 million tons. Roughly 25% of the ash is being recycled, and only 10% of it, i.e. 6 million tons/year, is being used as a mineral admixture in the cement and concrete industries. The ash utilization rates are much better in countries with relatively low ash production, such as Denmark, Sweden, and the Netherlands. For example, in the Netherlands, 98% of the 700,000 tons of fly ash produced in 1988 was utilized.

METALLURGICAL SLAGS

Iron and steel industries generate the largest amounts of slags; blast-furnace slag is produced during the production of pig iron from iron ore, and steel slags are produced during the conversion of pig iron to steel. Both slags are high in calcium (35–40% CaO). A cementitious granulated product results when these slags are rapidly cooled from the molten state. Copper and nickel slags are not cementitious because they are deficient in calcium. When rapidly cooled, they yield pozzolanic products.

According to Regourd (6), several slag granulation processes are now used in which air or water under pressure breaks down the liquid slag into small droplets which are solidified into grains of up to 4 mm size, mostly in a glassy state. This is called **granulated slag**. To avoid the formation of crystallization products, generally large quantities of water under pressure are used for making granulated slag. The wet slag has to be dried before it is ground for use as a mineral admixture. A semi-dry process, which produces **pelletized slag** (Fig. 2.1), has been developed in Canada. In this process, the molten slag is first cooled with a small amount of water, then flung into the air by a rotary drum. Several size fractions are obtained; larger pellets (4–15 mm) of expanded slag contain crystalline products and can be used as lightweight aggregate, whereas the smaller fraction (less than 4 mm) which is mostly glassy and cementitious can be ground for use as a mineral admixture.

Although the world production of iron blast-furnace slag is approximately 100 million tons/year, the utilization rate of the product as a cementitious material is poor because in many

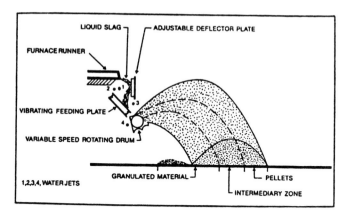

FIGURE 2.1   Diagram of a typical slag pelletizer. From reference 13.

countries, including the U.S., only a small proportion of the slag is granulated (for example, less than 10% of the total available blast-furnace slag is granulated in the U.S.). In addition to the U.S., China, France, Germany, India, and Japan are among the large producers of iron blast-furnace slag. In China and Japan the utilization rate of iron blast-furnace slag as a cementitious material is in the range of 35 to 60%.

SILICA FUME

Silica fume is a by-product of the ferro-silicon alloys and silicon metal industries. Silicon, ferro-silicon, and other alloys of silicon are produced in electric arc furnaces where quartz is reduced by carbon at very high temperatures. In the process, SiO vapors are produced which oxidize and condense in the form of very tiny spheres of non-crystalline silica (0.1 μm average diameter). The product, which is highly pozzolanic, is recovered by passing the outgoing flue gas through a baghouse filter. It seems that silica fume containing more than

78% SiO_2 in amorphous form is suitable for use in the cement and concrete industries. Due to extremely fine particle size and low bulk density, the handling and transportation of the material is generally carried out in the form of slurry or a pelletized product.

The current world production of silica fume appears to be about one million tons/year. Norway and the U.S.A, are among the major producers. In spite of several technical advantages, only a small percentage of the current supply of silica fume is being used as a mineral admixture in the cement and concrete industries. This may be due to handling difficulties and high cost of the material.

RICE-HUSK ASH

Worldwide, approximately 100 million tons/year of rice husks are available for disposal. As a renewable source of energy, the product is quite attractive as a substitute for fossil fuels for thermal power production. (The calorific value of rice husks is about 50% of coal.) A photograph of a power generation unit, using rice husks as fuel, is shown in Fig. 2.2. In addition to heat energy, the combustion of rice husks produces approximately 20% high-silica ash. This means that there is a potential for producing 20 million tons of rice-husk ash every year.

As crystalline silica is hazardous to human health, the burning temperature of a rice-husk ash combustion furnace must be controlled to keep silica in an amorphous state. Furthermore, the microporosity and the high surface area of the product also contribute to very high pozzolanic activity even when the material has not been ground to a fine particle size (such as silica fume and metakaolin). The engineering advantages from using the non-crystalline rice-husk ash as a mineral admixture in concrete, in particular a dramatic reduction in the permeability, have been recently reported (7,8); therefore it is expected that the material will soon find some use in the cement and concrete industries.

FIGURE 2.2 Photograph of a 25 MW power plant burning rice husks as fuel. The plant located at Williams, North California consumes 184,000 tons/year of waste rice husks. In the photograph, the husk storage silos are on the right, whereas a single-drum steam generator is on the left. A fabric filter is used to remove the rice-husk ash from the flue gases.

CHAPTER 3

Chemical and Physical Characteristics

In this chapter, chemical and physical characteristics of mineral admixture are described. It should be noted that, in general, differences in the chemical composition do not have any significant effect on the properties of mineral admixtures unless accompanied by significant mineralogical changes which, of course, also depend on the conditions of processing or formation. For instance, rapid cooling of a molten slag from high temperature produces more glassy phase, and therefore a more reactive material; the same slag when cooled slowly would essentially form crystalline solid phases with little or no reactivity. As discussed in Chapter 4, it is not so much the chemical composition but the mineralogical and granulometric characteristics, which determine how a mineral admixture would influence the important engineering properties of concrete, such as workability, strength, and impermeability. Therefore, in addition to the chemical composition, a brief review of the mineralogical and granulometric properties of various mineral admixtures is presented here.

CHEMICAL COMPOSITION

Typical oxide analyses of some natural pozzolans, fly ashes, slags, silica fume, rice-husk ash, and metakaolin are shown in Tables 3.1 to 3.4. With pozzolanic admixtures (Tables 3.1, 3.2, 3.4) it is commonly observed that the acidic constituents,

15

TABLE 3.1   Typical Oxide Analyses of Natural Pozzolans, [Ref. 9]

Source	Mass Percent						
	SiO_2	Al_2O_3	Fe_2O_3	CaO	MgO	Alkalies	Ignition Loss
Roman Tuff, Italy	44.7	18.9	10.1	10.3	4.4	6.7	4.4
Rheinish Trass, Germany	53.0	16.0	6.0	7.0	3.0	6.0	9.0
Santorin Earth, Greece	65.1	14.5	5.5	3.0	1.1	6.5	3.5
Jalisco Pumice, Mexico	68.7	14.8	2.3	—	0.5	9.3	5.6
Diatomaceous Earth, California	86.0	2.3	1.8	—	0.6	0.4	5.2

TABLE 3.2 Typical Oxide Analyses of North American Fly Ashes, [Ref. 10,11,12]

Source	Mass Percent							
	SiO_2	Al_2O_3	Fe_2O_3	CaO	MgO	Alkalies	SO_3	Ignition Loss
Bituminous, U.S.	55.10	21.10	5.20	6.70	1.60	2.97	0.50	0.60
Bituminous, U.S.	50.90	25.30	8.40	2.40	1.00	3.11	0.30	2.10
Bituminous, U.S.	52.20	27.40	9.20	4.40	1.00	0.80	0.45	3.50
Bituminous, Canada	48.00	21.50	10.60	6.70	0.96	1.42	0.52	6.89
Bituminous, Canada	47.10	23.00	20.40	1.21	1.17	3.70	0.67	2.88
Subbituminous, U.S.	38.40	13.00	20.60	14.60	1.40	2.44	3.30	1.60
Subbituminous, U.S.	36.00	19.80	5.00	27.20	4.90	2.12	3.15	0.40
Subbituminous, Canada	55.7	20.4	4.61	10.7	1.53	5.65	0.38	0.44
Lignite, U.S.	26.9	9.1	3.60	19.2	5.80	8.6	16.60	—
Lignite, Canada	44.5	21.1	3.38	12.9	3.1	7.05	7.81	0.82

TABLE 3.3   Typical Oxide Analyses of North American Blast-Furance Slags, [Ref. 13]

Source	Mass Percent						
	SiO$_2$	Al$_2$O$_3$	Fe$_2$O$_3$	CaO	MgO	Alkalies	Ignition Loss
Atlantic Cement, Sparrows Pt., Maryland, U.S.	33.17	10.8	0.63	41.60	12.5	0.85	0.54
Hamilton Steel Plant, Ontario, Canada	37.08	8.76	1.93	40.04	11.52	0.80	1.99
Algoma Slag, Sault Ste. Marie, Ontario, Canada	38.35	8.76	0.61	32.34	18.64	0.93	0.95

TABLE 3.4 Typical Oxide Analyses of Silica Fume, Rice-Husk Ash, and Metakaolin

Source	Mass Percent						
	SiO$_2$	Al$_2$O$_3$	Fe$_2$O$_3$	CaO	MgO	Alkalies	Ignition Loss
Silica fume from silicon metal industry (Ref. 14)	94.00	0.06	0.03	0.50	1.10	0.10	2.50
Silica fume from (75% Si) ferrosilicon alloy industry (Ref. 14)	90.00	1.00	2.9	0.10	0.20	2.20	2.70
Silica fume from (50% Si) ferrosilicon alloy industry (Ref. 14)	83.00	2.50	2.50	0.80	3.00	2.30	3.60
Rice-husk ash (Ref. 15)	92.15	0.41	0.21	0.41	0.45	2.39	2.77
Metakaolin (Ref. 4)	51.52	40.18	1.23	2.00	0.12	0.53	2.01

namely silica and alumina, vary widely not only between the admixture types but also within the same type. Similarly, in the case of cementitious admixtures, such as blast-furnace slag, the oxides of calcium and magnesium can show wide variations between the two slags (Table 3.3). However, as will be discussed in the next chapter, the chemical differences within a mineral admixture type or between the different types are by themselves not so important in determining the influence of these admixtures on properties of concrete.

As can be expected, among the natural pozzolans diatomaceous earth has a very high silica content (Table 3.1). Unaltered volcanic ashes, represented by Santorin Earth and Jalisco Pumice, also contain high content of silica, in addition to moderate amounts of alumina. Hydrothermal alteration products of volcanic materials, such as Roman Tuff and Rheinisch Trass contain relatively less silica and more CaO, MgO, and alkalies.

The chemical analyses of fly ashes (Table 3.2) show that, when compared to natural pozzolans, the fly ashes generally contain relatively less silica and more alumina. Fly ashes from the combustion of bituminous coals contain a low amount of calcium; on the other hand, subbituminous and lignite coal ashes typically contain more than 10% CaO in addition to higher amounts of alkalies and sulfates than usually present in the bituminous coal ashes. Lignite ashes are low in silica and alumina but usually contain large amounts of sulfates mostly in the form of sodium sulfate. Compared to natural pozzolans and fly ashes, blast-furnace slags are usually rich in calcium and magnesium oxides (Table 3.3).

The chemical composition of highly active pozzolanic materials is shown in Table 3.4. Except metakaolin they are characterized by a very high content of silica (e.g. more than 80% SiO_2); metakaolin may contain roughly equal proportions of SiO_2 and Al_2O_3 by mass. Depending upon the unburnt carbon present in rice-husk ash, the SiO_2 content of the commercially produced material generally ranges between 80 to 95%.

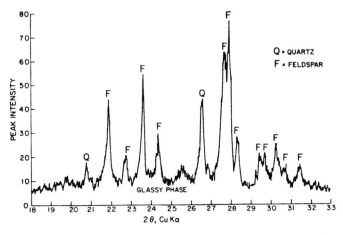

FIGURE. 3.1 X-ray diffraction analysis of Santorin earth. From reference 9.

MINERALOGICAL COMPOSITION

Natural pozzolans, composed of unaltered volcanic ashes, derive their pozzolanic activity from the aluminosilicate glass. Mineralogical analyses of typical samples of Santorin Earth from Greece, Jalisco pozzolan from Mexico, and Shirasu pozzolan from Japan showed 80%, 90%, and 95% aluminosilicate glass, respectively (9). The non-reactive crystalline components are composed of fragments of quartz, feldspar, and mica (Fig. 3.1). Mineralogical analyses of volcanic tuffs and trass also show significant amounts of quartz, feldspar, and clays in a glassy matrix that has undergone alteration to zeolitic minerals, such as analcite, chabazite, herschellite, phillipsite, and clinoptilolite. Rheinisch trass is reported to contain 50% of a glassy matrix that has been altered to zeolitic minerals. Diatomaceous earth is predominantly composed of opaline silica (hydrous, non-crystalline silica).

Mineralogical analyses of **fly ashes** typically show 50–90% glass, the chemical composition and the reactivity of glass in a fly ash being dependent on the calcium content of the fly ash. Low-calcium fly ashes from bituminous coals contain aluminosilicate glass which seems to be less reactive than the calcium aluminosilicate glass present in high-calcium fly ashes (Fig. 3.2). The crystalline minerals typically found in low-calcium fly ashes are quartz, mullite ($3Al_2O_3.2SiO_2$), sillimanite ($Al_2O_3.SiO_2$), hematite, and magnetite. These minerals do not possess any pozzolanic activity. The crystalline minerals typically found in high-calcium fly ashes are quartz, tricalcium aluminate ($3CaO \cdot Al_2O_3$), calcium aluminosulfate ($4CaO \cdot 3Al_2O_3 \cdot SO_3$), anhydrite ($CaSO_4$), free CaO, periclase (free MgO), and alkali sulfates. Except quartz and periclase, all the crystalline minerals present in the high-calcium fly ash react with water at ordinary temperature.

Mineralogical analyses of **granulated blast-furnace slag** samples show glass content ranging from 80 to 100%, the chemical composition of the glass being similar to melitite which is a solid solution phase between the gehlenite composition ($2CaO \cdot Al_2O_3 \cdot SiO_2$) and the akermanite composition ($2CaO \cdot MgO \cdot 2SiO_2$). It has been reported that the reactivity of the melilite glass generally goes up with the increasing ratio between the $2CaO \cdot Al_2O_3 \cdot SiO_2$ phase and the $2CaO \cdot MgO \cdot 2SiO_2$ phase. The non-reactive constituents in blast-furnace slag generally are the crystalline minerals gehlenite, akermanite, diopside ($CaO \cdot MgO \cdot 2SiO_2$) and merwinite ($3CaO \cdot MgO \cdot 2SiO_2$).

Silica fume, rice-husk ash, and metakaolin are highly reactive pozzolanic materials, which derive their lime-reactivity from the combination of two factors, namely totally non-crystalline structure and high surface area. Although they are comparable in regard to performance characteristics in concrete, their chemical and physical properties show some essen-

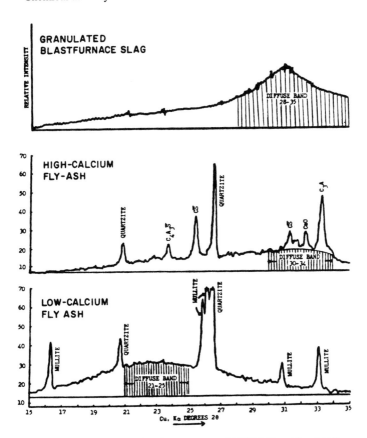

FIGURE. 3.2 X-ray diffraction patterns of ASTM Class F and C fly ashes and a granulated blast furnace slag. From reference 2.

tial differences and similarities between them. As far as the mineralogical character is concerned, all three are composed essentially of non-crystalline matter (Fig. 3.3). The non-crystalline phase in silica fume consists primarily of a disordered

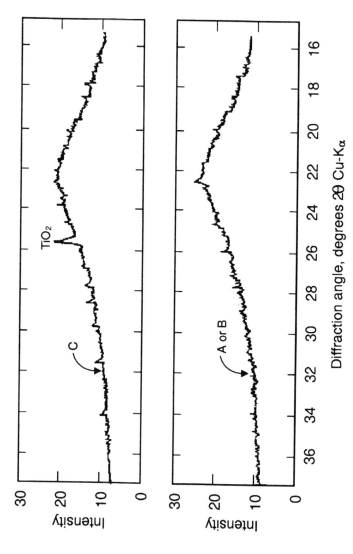

FIGURE 3.3 Typical x-ray diffraction patterns for non-crystalline pozzolans: A-silica fume, B-Rice-Husk Ash, C-Me-takaolin.

Si–O structure which is the product of solidification or condensation from a fused material. A similar disordered Si–O structure exists in the rice-husk ash burnt at temperatures less than about 700°C; however, the product is obtained by decomposition and sintering of opaline or hydrous silica without melting. Metakaolin consists of a non-crystalline aluminosilicate (Si–Al–O) phase that is obtained from a purified kaolin clay by a low-temperature burning process somewhat analogous to rice-husk ash production. Occasionally, a small amount of crystalline impurities may be present, viz. 1–2% cristobalite in rice-husk ash, and a similar amount of quartz, feldspar or titania in metakaolin.

PARTICLE CHARACTERISTICS

In general, the mechanisms by which mineral admixtures influence the properties of fresh and hardened concrete are dependent more on the size, shape, and texture of the particles than on the chemical composition. For instance, the water demand and workability are controlled by particle size distribution, packing effect, and smoothness of surface texture. The pozzolanic and cementitious properties, which govern the strength development and permeability of the blended cement system, are controlled by the mineralogical characteristics as well as the particle size and surface area of the mineral admixture.

Irrespective of whether a mineral admixture is an industrial by-product or is derived from natural sources, usually nothing can be done to alter its mineralogical characteristic. Therefore, the control of particle size distribution is the only practical method by which the pozzolanic or cementitious activity can be enhanced. ASTM C 618 Standard Specification for pozzolanic admixtures limits the large-size particles (i.e. particles retained on No. 325 mesh sieve or larger than 45 μm) to a maximum of 34%; commercially available materials seldom exceed 10% residue on No. 325 mesh. It is well known that particles larger than 45 μm show little or no reactivity under normal hydration conditions. From a study of the strength contribution potential of seven bituminous fly ashes from the U.S., Mehta (10) reported that the pozzolanic activity was directly proportional to the amount of particles under 10 μm. Similar results were reported for ground granulated blast-furnace slag by Wada and Igawa (16).

Surface area determination by air permeability methods, such as the Blaine apparatus, is commonly used in the portland cement industry as an easy substitute for particle size distribution. These methods are not suitable for very fine and microporous materials. With such materials, BET* nitrogen adsorption technique is preferred for surface area determination. When comparing the surface areas, it should be noted that no direct relationship exists between the Blaine and the BET surface areas. In addition to portland cement, typical particle size distribution plots for a fly ash and silica fume sample are shown in Fig. 3.4. In general, the glassy spheres in **fly ash** range from 1 to 100 μm, with an average size of about 20 μm. From a study of the U.S. fly ashes, it was reported (10) that, typically, more than 40% particles were under 10 μm and less than 15% particles were above 45 μm. The Blaine surface area ranged between 300 to 400 m^2/kg.

The typical microporous texture of volcanic ashes, diatomaceous earth, and rice-husk ash are shown by the scanning electron micrographs in Fig. 3.5. Fly ash and silica fume consist mostly of solid and spherical particles, although there is an order of magnitude difference in the size of spheres (Fig. 3.6). Particles of ground tuffs and slags generally exhibit a rough and compact texture.

The bituminous fly ashes look much cleaner under the microscope than subbituminous and lignite ashes (Fig. 3.7). This

---

*Brunauer, Emmet, and Taylor.

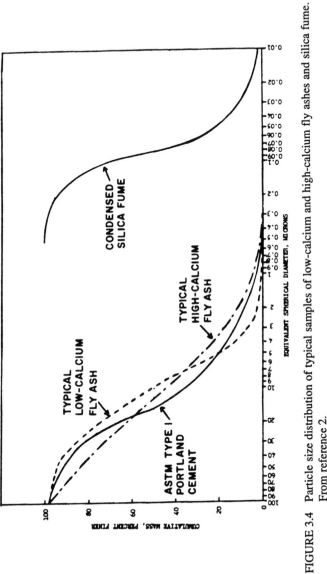

FIGURE 3.4   Particle size distribution of typical samples of low-calcium and high-calcium fly ashes and silica fume. From reference 2.

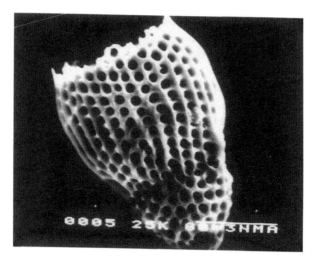

Diatomaceous

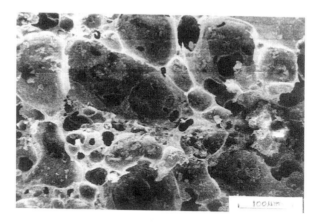

Volcanic Ash

FIGURE 3.5   Scanning electron micrograph showing the microporous nature of diatomaceous earth, volcanic ash, and
rice-husk ash.

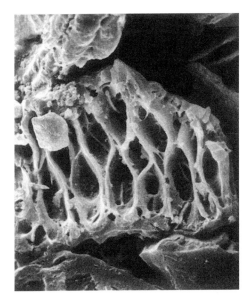

Rice-husk ash

FIGURE 3.5   (Continued)

is because alkali sulfates, which tend to crystallize on the surface of fly ash particles, are present in relatively insignificant
amount in bituminous fly ashes. Irregular masses of large-size
usually consist of either incompletely burnt carbonaceous matter or agglomeration of molten ash spheres. At times, some of
the spherical particles in fly ash are found to be hollow and either completely empty (called cenospheres) or filled with
smaller spheres (called plerospheres). A scanning electron micrograph of a broken plerosphere is also shown in Fig. 3.8. It
should be noted that a large volume of carbonaceous particles
or broken cenospheres, when present in a fly ash, would in-

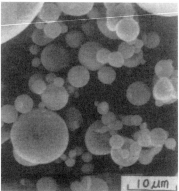

A

Fly ash

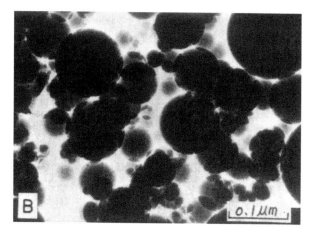

Silica fume

FIGURE 3.6   Scanning electron micrographs showing the size
difference between particles of fly ash and silica
fume.

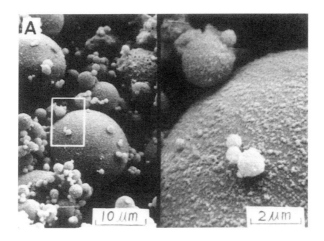

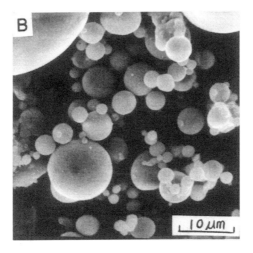

FIGURE 3.7   Scanning electron micrograph: (A)-ASTM Class C,
(B)-ASTM Class F fly ash.

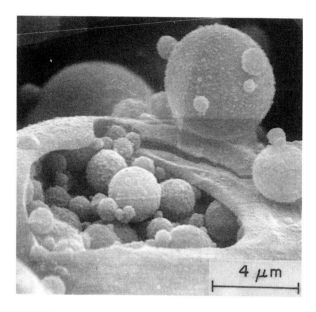

FIGURE 3.8   Scanning electron micrograph of a plerosphere containing cenospherical particles.

crease the surface area and therefore the water-reducing and
air-entraining admixture requirements.

Unlike fly ash, **granulated blast-furnace slag** and natural
pozzolans have to be ground to a desired particle size or surface area, depending upon the degree of activation needed and
economic considerations. It is observed (16) that with portland
cement-slag blends the slag particles < 10 μm contribute to
early strength development (up to 28–days); particles in the
10–45 μm range continue to hydrate beyond 28 days and contribute to later-age strength; and particles above 45 μm generally show a little or no activity. Typically, in order to obtain a
satisfactory strength performance in concrete the Blaine surface area of ground slag should range between 400 to 500

m^2/kg. The Canadian slag from the Sault Ste. Marie and the U.S. slag from the Sparrow Point steel plant (for chemical composition see Table 3.2) are typically ground to Blaine fineness of 450 m^2/kg and 500–550 m^2/kg, respectively (13). Also, both plants have closed-circuit grinding systems, which permit a closer control of the particle size distribution.

As stated earlier, the surface properties of microporous pozzolanic materials are better characterized by the BET nitrogen adsorption method. For instance, Mehta (9) reported studies on the pozzolanic activity of both Santorin Earth ground to 3800–15500 m^2/kg and a California diatomaceous earth ground to 12000 m^2/kg BET surface area. Typical BET surface areas of **silica fume** marketed for use as an admixture for concrete is approximately 20,000 m^2/kg. This is to be expected because the silica fume particles are extremely small, from 0.01 to 1 μm average diameter. Although **rice-husk ash** with high pozzolanic activity, is ground only to an average particle size of 6–10 μm, due to its microporous character and internal surface the nitrogen adsorption surface area values on the order of 40,000 to 100,000 m^2/kg are reported (7). Published information on the particle size characteristics of metakaolin is rather limited. The material commercially available in the U.S. is ground to an average particle size of about 1.5 μm (3). The material used in the French study (4) had a BET surface area value of 20,000 m^2/kg. It should be noted that pozzolanic materials with unusually high surface area not only exhibit excellent reactivity but also are able to impart stability and cohesiveness to concrete mixtures especially those that are prone to bleeding and segregation.

A classification of mineral admixtures, according to their pozzolanic and/or cementitious properties, is shown in Table 3.5. For the purposes of quick reference, a summary of the chemical, mineralogical, and particle characteristics is also included in this table.

TABLE 3.5   Classification, composition, and particle characteristics of mineral admixtures for concrete

Classification	Chemical and mineralogical composition	Particle characteristics
Cementitious and pozzolanic		
Granulated blast-furnace slag (cementitious)	Mostly silicate glass containing mainly calcium, magnesium, aluminum, and silica. Crystalline compounds of melilite group may be present in small quantity.	Unprocessed material is of sand size and contains 10–15% moisture. Before use it is dried and ground to particles less than 45 μm (usually about 500 m^2/kg Blaine). Particles have rough texture.
High-calcium fly ash (cementitious and pozzolanic)	Mostly silicate glass containing mainly calcium, magnesium, aluminum, and alkalies. the small quantity of crystalline matter present generally consists of quartz and C$_3$A; free lime and periclase may be present; C$\overline{S}$ and C$_4$A$_3\overline{S}$ may be present in the case of high-sulfur coals. Unburnt carbon is usually less than 2%	Powder corresponding to 10–15% particles larger than 45 μm (usually 300–400 m^2/kg Blaine). Most particles are solid spheres less than 20 μm in diameter. Particle surface is generally smooth but not as clean as in low-calcium fly ash.
Highly-active pozzolans		
Condensed silica fume	Consists essentially of pure silica in noncrystalline form.	Extremely fine powder consisting of solid spheres of 0.1 μm average diameter (about 20 m^2/g surface area by nitrogen adsorption.
Rice husk ash	Consists essentially of pure silica in noncrystalline form.	Particles are generally less than 45 μm but they are highly cellular (about 60 m^2/g surface area by nitrogen adsorption.
Normal pozzolans		
Low-calcium fly ash	Mostly silicate glass containing aluminum, iron, and alkalies. The small quantity of crystalline matter present generally consists of quartz, mullite, sillimanite, hematite, and magnetite.	Powder corresponding to 15–30% particles larger than 45 μm (usually 200–300 m^2/kg Blaine). Most particles are solid spheres with average diameter 20 μm. Cenospheres and plerospheres may be present.
Natural materials	Besides aluminosilicate glass, natural pozzolans contain quartz, feldspar, and mica.	Particles are ground to mostly under 45 μm and have rough texture.
Weak pozzolans		
Slowly cooled blast-furnace slag, bottom ash, boiler slag, field burnt rice husk ash	Consists essentially of crystalline silicate materials, and only a small amount of noncrystalline matter.	The material must be pulverized to very fine particle size in order to develop some pozzolanic activity. Ground particles are rough in texture.

CHAPTER 4

Mechanisms by which Mineral Admixtures Improve Properties of Concrete

By using mineral admixtures in concrete it is possible to have a favorable influence on many properties through either purely physical effects associated with the presence of very fine particles or physico-chemical effects associated with pozzolanic and cementitious reactions, which result in pore-size reduction and grain-size reduction phenomena, as discussed below. Among the properties that are favorably affected are the rheological behavior of fresh concrete mixtures, and the strength and durability of hardened concrete. Resistance to chemical attacks and thermal cracking are the two aspects of concrete durability that can be improved significantly by the incorporation of mineral admixtures.

Field experience with mineral admixtures shows that the above-mentioned potential benefits expected from their use are not always realized. Therefore, it is necessary to have a proper understanding of the mechanisms by which mineral admixtures improve properties of concrete.

WATER-REDUCING EFFECT OF MINERAL ADMIXTURES

Many researchers have reported that partial replacement of

35

portland cement by some mineral admixtures, such as fly ash, in a mortar or concrete reduces the water requirement to obtain a given consistency. The phenomenon is generally attributed to spherical shape and smooth surface of fly ash particles. Helmuth (17) does not consider this explanation as adequate because the angular and rough-textured particles of ground granulated blast-furnace slag are also known to have a water-reducing effect on portland cement mortar and concrete. The author believes that the water reduction caused by fly ash is the result of an adsorption-dispersion mechanisms which is similar to the action of water-reducing chemical admixtures. Very fine particles of fly ash get adsorbed on the oppositely-charged surface of cement particles and prevent them from flocculation. The cement particles are thus effectively dispersed and will not trap large amounts of water, which means that the system will have a reduced water requirement for flow.

. In addition to the above mechanisms, particle packing effect is also responsible for water reduction. Note that portland cement particles are mostly in the size range of 1 to 50 μm. Therefore, physical effect of particle packing by the microfine particles of a mineral admixture will reduce the void space and correspondingly the water requirement for plasticizing the system. Fig. 4.1 shows an illustration as to how this mechanism would operate in the portland cement-silica fume system. Theoretically, very fine particles of cement should also function in the same way. But they tend to dissolve rapidly in water, and are therefore ineffective as a space filler. However, any microfine particles that are relatively inert should be effective. Also, with regard to microfine fillers which contain a very large proportion of very fine particles, such as silica fume, it should be obvious that the filler particles themselves must be dispersed with the aid of a plasticizing agent before any benefit from the particle packing effect can materialize.

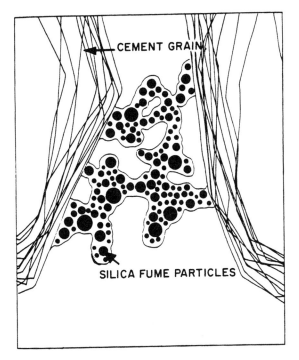

FIGURE 4.1   Mechanism of bleeding reduction in cement paste by silica fume addition (Courtesy of L. Hjorth, Aalborg Cement Co., Denmark).

RHEOLOGICAL BEHAVIOR OF CONCRETE

In general, due to better particle packing and less water requirement, fresh concrete containing mineral admixtures shows a reduced tendency for segregation and bleeding. Consequently, improved characteristics result with regard to cohesiveness, workability, pumpability, and finishing. Very fine mineral admixtures, such as silica fume, produce certain negative effects caused by close particle packing. When used in amounts exceeding 3–4% by mass of cement, they impart stickiness and a thixotropic tendency to fresh concrete.

STRENGTH, WATER-TIGHTNESS, AND CHEMICAL DURABILITY

Too much mixing-water in a concrete mixture is probably the most important cause for many problems that are encountered with concrete, both in the fresh and hardened state. There are two reasons why typical concrete mixtures contain too much mixing-water. Firstly, the water demand and workability of a concrete mixture are influenced greatly by particle size distribution, particle packing effect, and voids present in the solid system. Typical concrete mixtures made with locally available materials do not have an optimum particle size distribution, and this accounts for the undesirably high water requirement to achieve certain workability. Secondly, to plasticize a cement paste for achieving a satisfactory consistency, much larger amounts of water than necessary for hydration of cement have to be used because portland cement particles, due to the presence of electronic charges on the surface, tend to form flocs that trap large volumes of the mixing water (Fig. 4.2).

When a concrete mixture is consolidated after placement, along with entrapped air a part of the mixing-water is also released. As water has low density, it tends to travel to the surface of concrete. However, not all of this "bleed water" is able to find its way to the surface. Due to the wall effect of coarse aggregate particles, some of it accumulates in the vicinity of aggregate surfaces, causing a heterogeneous distribution of water in the system. Obviously, the areas with high w/c have more available space which permits the formation of a highly porous hydration product composed mostly of large crystals of

FIGURE 4.2 Diagrammatic representation of the flocculated cement paste system.

calcium hydroxide and ettringite. Microcracks due to stress are readily formed through this product (Fig. 4.3), because it is more porous and much weaker than the areas with lower w/c.

It has been suggested that the microcracks in the interfacial transition zone play an important part in determining not only strength but also the permeability and durability of concrete exposed to severe environmental conditions. This is because the rate of fluid transport in concrete can be much larger by percolation through an interconnected network of microcracks than by diffusion or capillary suction. The heterogeneities in the microstructure of the hydrated cement paste, especially the existence of large pores and large-crystalline products in the transition zone, can be greatly reduced by the introduction of fine particles of pozzolanic or cementitious admixtures. With the progress of the pozzolanic and cementitious reactions, there is a gradual decrease in the size of pores and crystalline hydration products (18). Initially, these two effects can result from closer particle packing phenomenon alone. Note that at the interfacial zone, the conditions for the pozzolanic reaction are more favorable because the finer particles of both cement

FIGURE 4.3 Intercrystalline cracking in calcium hydroxide present in the transition zone between coarse aggregate and cement paste.

and mineral admixtures tend to exist here in a relatively high w/c paste. In short, a combination of both physical and chemical effects accounts for the high strength of the transition zone in concrete containing mineral admixtures.

A strong transition zone is essential for producing a high-performance concrete that is characterized by several properties including high strength, high water-tightness, and high durability. As interconnected microcracks in the transition zone are implicated in the loss of water-tightness and problems associated with durability of concrete exposed to aggressive chemicals, it should be clear how the incorporation of mineral admixtures into concrete helps the production of high-performance products.

RESISTANCE TO THERMAL CRACKING

Thermal cracking is a matter of serious concern in the design of mass concrete structures. It is generally assumed that this is not a problem with structures of moderate thickness, viz. 1 m or less. However, cases of thermal cracking are reported even from moderate-size structures as concrete mixtures with high cement content tend to develop excessive heat during curing. This is because, first, the physical-chemical composition of ordinary portland cement today is such that it has high heat of hydration compared with normal portland cement available 40 years ago. Secondly, high-early strength requirements in modern construction practice are satisfied either by increasing the cement content of a concrete mixture or by heat curing. Thirdly, there is considerable construction activity now in the hot-arid areas of the world where concrete temperature in excess of 60°C are not uncommon.

According to Idorn (19), petrographic studies of concrete cores from many recently built structures have shown severe microcracking during early curing. A probable cause of this microcracking is thermal stress generated by rapid cooling of hot concrete. This is why the Danish Highway authority has limited the temperature in the concrete interior to a maximum of 60°C and has also specified a cooling rate such that the temperature difference between the surface and the interior shall not exceed 20°C. American Concrete Institute has a similar recommended practice for mass concrete construction. Idorn (19) cautions that the thermal cracks may not be visible in the beginning but they would gradually propagate to larger and wider cracks due to stresses caused by weathering and loading effects.

Moderately-active mineral admixtures, such as natural pozzolans and low-calcium fly ashes do not show any significant chemical interaction with cement hydration products during the first week of hydration; the slow pozzolanic and cementitious reactions generally begin thereafter. Early-age cracking

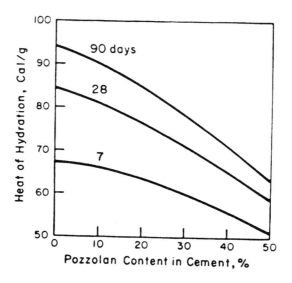

FIGURE 4.4   Effect of substituting an Italian natural pozzolan on
the heat of hydration. From reference 22.

from thermal stresses generated by rapid cooling of hot concrete can therefore be controlled by a partial substitution of cement with a moderately-active pozzolan. In fact, this was the
consideration underlying the first large-scale use of fly ash in
the U.S., namely the construction project for the Hungry Horse
Dam in 1947 (20). It has been estimated that the contribution
of low-calcium fly ash to early-age heat generation ranges
from 15 to 30% of that of an equivalent mass of portland cement (21). Massazza and Costa's data (22) on the effect of substitution of an Italian natural pozzolan on the heat of hydration
are shown in Fig. 4.4. Portland cement-slag blends containing
50% mass of a granulated blast-furnace slag show similar results when the slag has not been ground to a high fineness.

It should be obvious that with blended cements containing
relatively small amounts (i.e. 5–10% by mass of cement) of

highly active pozzolanic materials, only small reduction in heat of hydration will be obtained. The development of a high heat of hydration and consequently a high concrete temperature becomes unavoidable, when the selection of materials and proportions is dictated by consideration of high strength and very low permeability at early ages.

CHAPTER 5

Effect of Mineral Admixtures on Properties of Fresh Concrete

WATER DEMAND

The water demand of the concrete incorporating a mineral admixture depends mainly on physical characteristics of the admixture as discussed in Chapter 4. For instance, the small size and the essentially spherical form of low-calcium fly ash particles have been credited with influencing the rheological properties of cement paste; this causes a reduction in the amount of water required for a given degree of workability compared to that required for an equivalent paste without fly ash. In this respect, as noted by Davis et al. (23), fly ash differs from natural pozzolans which usually increase the water requirement of concrete mixtures due to the microporous nature and angular shape of their particles.

Compton and MacInnis (24) reported that a concrete made by substituting 30% of the cement with a Canadian fly ash required 7% less water than was required for a control concrete of equal slump. Pasko and Larson (25) examined the amount of water required to maintain a nominal 60–mm slump in concrete mixtures with partial replacement of cement by fly ash. They found that the water requirement was reduced by 7.2% in a mixture in which 30% fly ash replaced 20% cement. During investigations of the concrete materials for construction of the

45

South Saskatchewan River Dam, Price (26) found that water requirement was not increased when additions of fly ash were made to concrete proportioned with fixed cement contents. The resulting concrete had a lower ratio of water-to-total cementitious material, yet workability and cohesiveness of the mixtures were improved.

Brink and Halstead (27) reported that some fly ashes reduced the water requirement of test mortars, whereas others generally of higher carbon content showed increased water requirement above that of the control mortars.

Welsh and Burton (28) reported loss of slump and flow for concretes made with some Australian fly ashes used to partially replace cement, when water content was maintained constant. Rehsi (29) reported that experience with a number of Indian fly ashes showed that all those examined increased the water requirement of concrete. In general, poor water demand characteristics have been found for fly ashes from older power plants where high carbon and coarse particle size are prevalent.

According to Owens (30), the major factor influencing the effect of fly ash on the workability of concrete is the proportion of coarse material (>45 μm) in the ash. He has shown that, for example, substitution of 50% by mass of the cement with fine particulate fly ash can reduce the water demand by 25%; a similar substitution using ash with 50% of the material greater than 45 μm had no effect on water demand. The general effect of coarse fly ash particles on the water demand is illustrated in Fig. 5.1.

Brown (31) examined the workability of four concrete mixtures of different water-to-cement ratios in which fly ash was substituted for cement on an equal volume basis. Slump, time of flow as measured in Vebe consistometer test, and compacting factor were measured for each mixture. It was found that both slump and workability increased with increasing fly ash substitution. The changes were found to depend upon the level of ash substitution (small additions sometimes being ineffectual) and on the water content. An empirical estimate was made

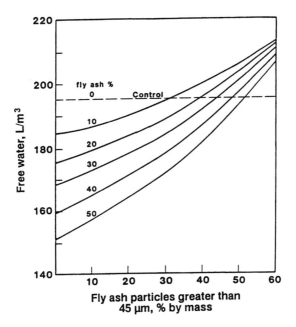

FIGURE 5.1 Influence of coarse particulate content of fly ash on the water required for equal workability in concrete. From reference 30.

which indicated that for each 10% of ash substituted for cement, the compacting factor changed to the same degree as it would by increasing the water content of the mixture by 3 to 4%.

In another series of experiments, Brown determined the effect of fly ash substitution for equal volumes of aggregate or sand in one concrete, keeping all other mixture proportions (and the aggregate grading) constant. The test concrete was modified by replacing either 10, 20, or 40% of the volume of sand by ash, or 10, 20, or 40% of the volume of the total aggregate by ash. The replacement of 40% of the total aggregate gave a mixture that was unworkable. It was concluded that when fly ash was substituted for sand or total aggregate, workability increased to reach a maximum value at about 8% ash by volume of the aggregate. Further substitution caused rapid decreases in workability.

An investigation by Carette and Malhotra (32) has also shown that water reduction does not always accompany the inclusion of fly ash, even with ash of otherwise acceptable properties obtained from modern power plants. Of the eleven fly ash concrete mixtures for which data are presented in Tables 5.1 and 5.1A, nine caused significant increases in slump at constant water content.

Silica Fume

Owing to its spherical, very small-sized particles, silica fume should be able to fill the space between large grains of cement, leading to a decrease in water demand in silica-fume concrete. However, the high specific surface area of silica fume particles tends to increase the water demand, giving a net effect of increased water requirement compared to a portland cement concrete with the same level of consistency. Normal and high-range water reducers are generally used to advantage with silica fume concretes in order to reduce the water demand.

Investigations dealing with the water demand to reach a given slump for different percentages of silica fume have been reported by Johansen (33) and Løland and Hustad (34). For very lean concrete (cement content = 100 kg/m^3), it was found that the water demand decreased with the addition of silica fume. Similar results have been reported by Aitcin et al. (35) for lean concrete mixtures with less than 10 per cent silica fume.

In concrete mixtures with a cement content of more than 250 kg/m^3 and without water reducing admixtures, the water demand increases with the addition of the silica fume. Sellevold and Radjy (36) analyzed the published data by Dagestad

(37), and concluded that in 1m³ of concrete the increased water demand per kilogram of silica fume added was about one litre. However, the use of water-reducing admixtures significantly reduced the water demand of the silica fume concretes. The use of dry lignosulphonate powder (0.2 to 0.4%) by weight of cement was sufficient to make the water demand of the concrete incorporating 10 per cent silica fume equal to the control concrete.

Slag

Meusel and Rose (38) have reported that with an increasing proportion of slag as a cementitious material in concrete, there is an increase in slump thereby indicating lower water demand for slag-concrete; however, exceptions have been reported.

Natural Pozzolans

Most natural pozzolans, when incorporated into concrete, tend to increase the water requirement of concrete because of their microporous character and high surface area. The higher water demand of a concrete mixture containing a natural pozzolan may not necessarily impair the strength because a part of the mixing water will be absorbed by the pozzolan, and thus, will not cause a corresponding increase in the porosity of the hardened concrete. According to Nicolaidis (39), this absorbed water would later be available for the pozzolanic reaction.

AIR ENTRAINMENT

When compared with the control concrete mixture, all mineral admixtures cause an increase in the dosage of an air-entraining admixture to obtain a given amount of entrained air. In the case of silica fume, slag and natural pozzolans, this is

TABLE 5.1   Mixture proportions of concretes incorporating some Canadian fly ashes*

| Mix No.*** | Batch quantities, kg/m³ | | Aggregate | | AEA, mL/m³ | W/(C + F)** | Cement replacement by fly ash % by mass |
	Cement	Fly Ash	Fine	Coarse			
Control	295	0	782	1082	170	0.50	0
F1	236	59	780	1077	320	0.50	20
F2	237	59	782	1080	200	0.50	20
F3	237	59	786	1088	200	0.50	20
F4	238	59	792	1094	160	0.50	20
F5	237	59	782	1080	690	0.50	20
F6	238	59	784	1082	660	0.50	20
F7	239	59	780	1077	370	0.50	20
F8	236	59	775	1069	230	0.50	20
F9	236	59	775	1070	240	0.50	20
F10	237	59	781	1079	290	0.50	20
F11	237	59	782	1080	150	0.50	20

*From reference 32.
**Water-to-(cement + fly ash) ratio by mass.

TABLE 5.1A   Proportions of fresh concrete mixtures from Table 5.1.

Mixture No.	Slump mm	Air %	Unit wt. kg/m³	Bleeding, %	Setting time, h:min	
					Initial	Final
Control	70	6.4	2320	2.9	4:10	6:00
F1	100	6.2	2300	3.1	4:50	8:00
F2	105	6.2	2310	4.6	7:15	10:15
F3	100	6.2	2310	5.1	5:20	8:10
F4	110	6.3	2320	4.3	6:20	8:25
F5	65	6.4	2310	2.7	5:15	8:55
F6	75	6.5	2300	2.6	4:30	6:50
F7	100	6.1	2300	2.9	4:15	6:20
F8	115	6.2	2300	5.6	5:10	7:30
F9	100	6.4	2280	4.4	5:25	9:00
F10	130	6.5	2290	2.5	4:45	7:00
F11	140	6.6	2290	0.6	4:00	6:05

mainly due to the high surface area of the material; in the case of fly ash, it is mainly due to the presence of carbon particles in the ash as the air-entraining admixture tends to get adsorbed on the surface of the carbon particles. Gebler and Klieger (40) offered the following conclusion relevant to air entrainment in fresh concrete:

"As the organic matter content, carbon content and loss-on-ignition of fly ash increase, the air-entraining admixture requirement increases as does the loss of air in plastic concrete."

Carette and Malhotra (32) have published considerable data on air-entrained concrete incorporating both ASTM Class F or Class C fly ashes. In one investigation, the air-entraining admixture dosage needed to entrain 6.4% air in concrete increased from 170 mL/m^3 for control concrete to 690 mL/m^3 for concrete incorporating 60% fly ash. This is an extreme case; in general the increase was considerably less (Table 5.1).

In another investigation by Malhotra (41), the admixture dosage needed to entrain about 5% air increased from 177 mL/m^3 for the control concrete to 562 mL/m^3 for the concrete mixture incorporating 65% slag. The water-to-(cement+slag) ratio was 0.30. At higher water-to-(cement+slag) ratios, the increase was not as marked as at the lower ratios.

Because of the extremely high surface area of the silica fume, the dosage of air-entraining admixture required to produce a certain volume of air in silica-fume concrete increases considerably with increasing silica fume dosage (Fig. 5.2). It has been reported that entrainment of more than 5% air is difficult in concrete incorporating high amounts of silica fume, even in the presence of a superplasticizer (42). Also, it was reported that the air-entraining admixture dosage increased significantly with silica fume content at only low water-to-cementitious materials ratio when comparing concretes made with various water-to-cementitious materials ratios.

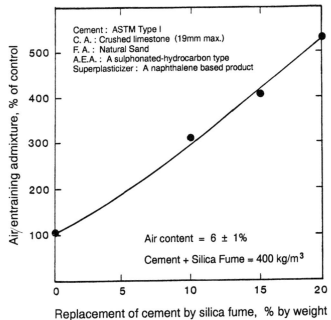

FIGURE 5.2   Relation between air-entraining admixture demand and dosage of silica fume. From reference 42.

Rice-Husk Ash

Zhang and Malhotra* have shown that at the cement replacement level of 10%, rice-husk ash concrete required more superplasticizer and more air-entraining admixture compared with the control portland cement concrete in order to obtain the same slump and air content, respectively. They attributed this to the high specific surface and high carbon content of the rice-husk ash used.

*CANMET Division Report MSL 95–007(OP&J)

BLEEDING

Fly Ash

The bleeding of fly ash concrete depends on the manner in which fly ash is used. When fly ash is used as direct replacement for cement (on weight basis), the volume of fly ash is more than the volume of cement replaced, thus resulting in more paste volume, which should reduce bleeding. This is not always so because the above benefit is offset by the water-reducing nature of the fly ash particles, which would result in bleeding when the water content of the concrete mixture is not reduced. In one study by Carette and Malhotra (32), it was found that all except two of the eleven ashes examined increased the bleeding (Table 5.1).

Concrete using fly ash generally shows reduced segregation and bleeding, and is therefore suitable for placement by pumping (43). Copeland (44) reported that in the field, the use of fly ash was found to reduce bleeding in concretes made from aggregates known to produce harsh mixtures normally prone to bleeding. Johnson (45) reported that most concrete made in the Cape Town (South Africa) area suffers from excessive bleeding due to lack of fines in the locally available dune sands. He added that the problem can be overcome by increasing the overall paste volume by using fly ash in concrete.

Blast-Furnace Slag

Few published data are available on the bleeding of slag concretes. Slags are generally ground to a higher fineness than normal portland cement, and therefore, a given mass of slag has a higher surface area than the corresponding mass of portland cement. As the bleeding of concrete is governed by the ratio of the surface area of solids to the volume of water, in all likelihood the bleeding of slag concrete will be lower than that of the corresponding control concrete. The slags now available in Canada and the U.S.A. have fineness, as measured by the Blaine method, greater than 4000 cm²/g compared with that of about 3000 cm²/g for portland cement. Thus, with concrete in which a given mass of portland cement is replaced by an equivalent mass of slag, bleeding should not be a problem. Figs. 5.3 and 5.4 show comparative data on the

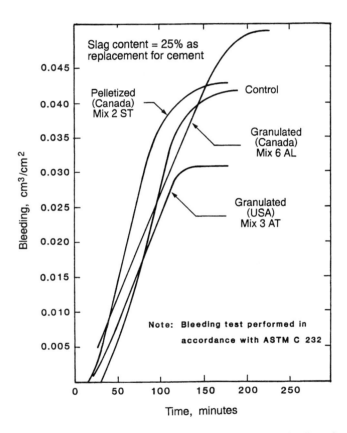

FIGURE 5.3 Comparative data on rate of bleeding and bleeding of concrete incorporating slags at 25% cement replacement level. From reference 46.

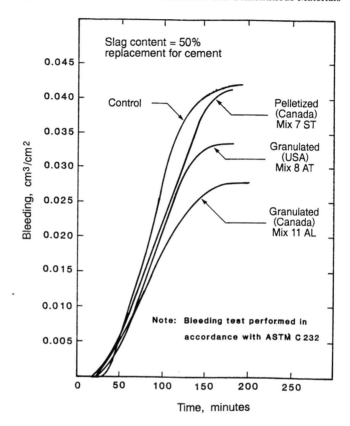

FIGURE 5.4   Comparative data on rate of bleeding and bleeding of
concrete incorporating slags at 50% cement replacement levels. From reference 46.

rate of bleeding of portland cement concrete, and bleeding of
concrete incorporating different percentages of a slag (46).

Silica Fume

Bleeding of silica fume concrete is considerably lower than
that of plain portland cement concrete. As illustrated in Fig.

4.1 in Chapter 4, the extremely fine silica-fume particles distribute themselves between the cement particles, reducing the channels for bleeding, thus allowing very little free water to rise to the surface of freshly consolidated concrete (33,34,47,48). Fig. 5.5 shows a significant reduction in bleeding with increasing silica fume replacement (49).

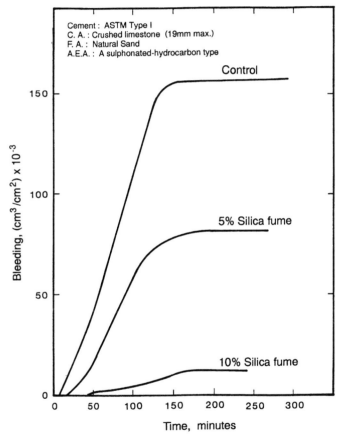

FIGURE 5.5   Bleeding rate of control and silica fume concrete. From reference 49.

Natural Pozzolans

With blended cements containing natural pozzolans, many investigators have observed that the rate of bleeding is reduced considerably. For example, using the ASTM Method C 243–55, Nicolaidis (39) found bleeding rates of 127×10^{-6} and 84×10^{-6} cm^3/cm^2 per sec. for the portland cement and for the portland-pozzolan cement (containing 20% Santorin earth), respectively. Even more impressive reduction in bleeding was obtained by Davis and Klein (50) who used finely ground diatomaceous earth from Lompoc, California, in a lean concrete mixture. The one-hour bleeding in the reference concrete (without the diatomaceous earth) amounted to 22% of total water present. When 11% of the cement by weight was replaced with the diatomaceous earth, the bleed water was only 2% of the total mixing water.

The reduction in bleeding results partly from the interference provided by the finely pulverized particles of the pozzolan to the water flow channels in a freshly consolidated concrete mixture, and partly from adsorption by the microporous pozzolanic material. It is believed now that the control of internal bleeding in concrete plays an important role in determining the strength of the transition zone between aggregate and cement paste, and therefore, the mechanical properties of concrete (18).

Rice-Husk Ash

Zhang and Malhotra* have shown that the bleeding of rice-husk ash concrete, at a water-to-(cement + rice husk ash) ratio of 0.40 and incorporating 10% of the ash as replacement for portland cement, was negligible.

---

*CANMET Division Report MSL 95–007 (OP&J)

SETTING TIME

There seems to be a general agreement in the published literature that low-calcium fly ash, slag and natural pozzolans exercise some retarding influence on the setting time of concrete, whereas the incorporation of up to 10 percent silica fume by weight of cement, does not significantly affect the setting characteristics of concrete.

The data in Table 5.1A, from the experiments conducted by Carette and Malhotra (32), show that all except one of the eleven ashes examined significantly increased both the initial and final setting times. Fly ashes with CaO content ranging from 1.4 to 13.0% were included in this study.

Lane and Best (51) confirmed that fly ash generally slowed the setting of concrete. They concluded that the observed retardation was affected by the proportion, fineness, and chemical composition of the ash; however, the cement fineness, the water content of the paste and the ambient temperature were considered to have a much greater effect. Davis et al. (50) concluded that fly ash-cement mixtures set more slowly than corresponding cements, but that the setting times were within the usual specified limits.

Research on fly ash concretes incorporating larges volumes of fly ash has indicated that the initial and final setting times of the concrete, though longer than that of control concrete, were found to be adequate for practical applications.

Mailvaganam et al (53) examined properties of fresh and hardened concretes made with a low-calcium fly ash in the presence of various other admixtures. Concrete mixed at 5°C showed retardation of setting time in excess of 10 h, regardless of fly ash content; concrete mixed at 20°C, and containing 30% of fly ash (by weight of cement) showed that setting time was extended by approximately 1 to 1.75h.

Dodson (54) has drawn the following conclusions from his work on the effect of fly ash on the setting time of concrete:

Initial and final setting times are influenced by the water-to-cement ratio and the total cement content of the concrete. The combined influence of these parameters can be expressed in terms of the Omega Index Factor (OIF) where OIF = cement factor ÷ W/C.

The relationship between setting times and OIF of concretes made from different cements, while each following the same general form, differs from one another considerably.

Whereas measured setting times for concretes containing low-calcium fly ashes are extended, the extended setting time is ascribed to the secondary influences of dilution of the portland cement content.

A high-calcium fly ash examined (25.5% CaO) had an accelerating influence on the setting time. This may be ascribed to the inherent cementitious characteristics of this particular ash, a property which is exhibited by many subbituminous fly ashes containing significant content of C_3A, $C_4A_3\overline{S}$, alkali-sulfates, and free CaO.

The observation of reduced setting time in the presence of high-calcium fly ash is by no means general. Ramakrishnan et al. (55) studied the setting time of mortars from concretes made with ASTM Type I and III cements both with and without a high-calcium fly ash (20.1% CaO). The data from their study are shown in Fig. 5.6. It is clear that in the presence of high-calcium fly ash, the setting time was retarded for both types of cements with the effect being minimal for Type III cement.

Silica Fume

Investigations show that ordinary concrete mixtures (with 250 to 300 kg/m³ cement) incorporating small amounts of silica fume, up to 10% by weight of cement, exhibit no significant difference in setting time compared to conventional concrete (56). As silica fume is invariably used in combina-

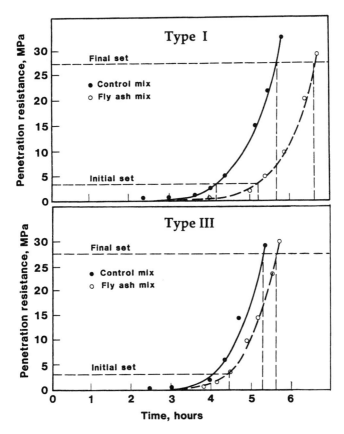

FIGURE 5.6 Comparison of setting time of control and fly ash concretes. From reference 55.

tion with water reducers and superplasticizers, the effect of silica fume on setting time of concrete is masked by these admixtures. The investigation by Bilodeau (49) has shown that the addition of 5 to 10% silica fume to the superplasticized and non-superplasticized concretes had a negligible effect on

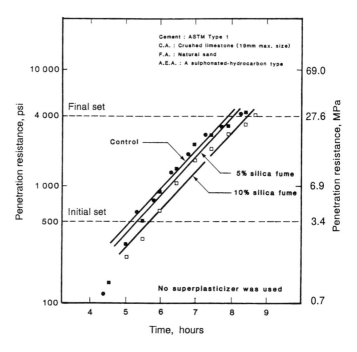

FIGURE 5.7   Penetration resistance of control and non superplasticized silica fume concrete. From reference 49.

the setting time of concrete (Fig. 5.7). However, in a concrete with water-to-cementitious materials ratio of 0.40 and 15% silica fume, there was a noticeable delay in the setting time. Due to the high silica fume content in the concrete, the high dosage of the superplasticizer could have contributed to this phenomenon (Fig. 5.8).

Blast-Furnace Slag

The incorporation of slag as a replacement for portland cement in concrete normally results in increased setting time of

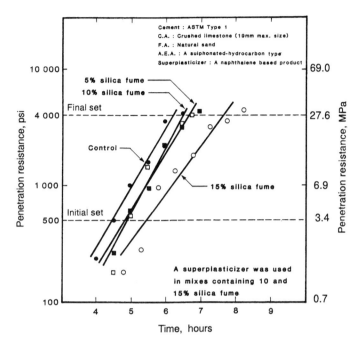

FIGURE 5.8   Penetration resistance of control and superplasticized
silica fume concrete. From reference 49.

concrete. Final setting time can be delayed up to several hours
depending upon the ambient temperature, concrete temperature, and mixture proportions. At temperatures lower than
23°C, considerable retardation in setting time can be expected
for slag concretes compared with control concrete; this has
serious implications in winter concreting. At high temperatures (>30°C), there is little or no change in setting time of
slag concrete as compared with that of control concrete (57).
Data by Hogan and Meusel (57) on initial and final setting
times for concrete incorporating granulated slag are shown in
Table 5.2.

TABLE 5.2 Data on time of setting for air-entrained concrete incorporating granulated slag*

Properties	Slag content, %				Control, no slag	Slag content, %		
	Control, no slag	40	50	65		40	50	60
Water/(cement + slag)	0.40	0.40	0.40	0.40	0.55	0.55	0.55	0.55
Cement factor (cement and slag), kg/m³	413	435	419	408	272	290	245	301
Fine aggregate/coarse aggregate	33/67	33/67	33/67	33/67	46/54	46/54	46/54	46/54
Air content, %	5.4	3.4	4.5	5.0	4.3	3.5	6.0	4.9
Unit weight, kg/m³	2330	2350	2320	2310	2345	2355	2310	2295
Slump, mm	75	75	95	90	70	15	50	75
Air-entraining admixture mL/kg cement	4.4	4.4	9.0	9.5	8.4	4.6	8.2	8.2
Initial set at 21.1°C (70°F), h:min	4:06	4:02	4:31	4:30	4:32	5:02	5:10	5:21
Final set at 21.1°C (70°F), h:min	5:34	5:40	6:29	7:04	7:03	6:40	8:10	8:09
Initial set at 32.2°C (90°F), h:min	3:30	—	3:45	—	—	—	—	—
Final set at 32.2°C (90°F), h:min	4:10	—	4:50	—	—	—	—	—

*From reference 57.

Malhotra's (46) data on the time of setting of concrete incorporating a pelletized slag and two different granulated slags are shown in Table 5.3. In this study, at 25% cement replacement, there was no significant increase in the initial setting time, whereas the increase in the final setting time ranged from 16 to 101 min. At 50% cement replacement, the increase in the initial setting time of the concretes ranged from 17 to 80 min., whereas the increases in the final setting time ranged from 93 to 192 min.

Natural Pozzolans

The addition of a natural pozzolan to portland cement results in set retardation partly because of the dilution effect (i.e., the dilution of the most active ingredient, which is portland cement) and partly because of the increased water requirement for making the cement paste of normal consistency. For instance, Nicolaidis (39) reported that, compared to the setting times for neat portland cement, both the initial and final setting times of a blended portland cement containing 20% Santorin earth were increased by 20 min. The water requirements to produce a paste of normal consistency were 24.5 and 26.5% for the portland cement and the blended portland-pozzolan cement, respectively.

Rice-Husk Ash

According to Zhang and Malhotra* the final setting times of the portland cement concrete incorporating 10% rice-husk ash were only slightly longer than that of the control concrete.

---

*CANMET Division Report MSL 95–007 (OP&J)

TABLE 5.3   Bleeding and setting-time of slag concrete: comparative data on three different slags*

Mixture no.	Type and source of slag	Blaine fineness of slag, cm²/g	Cement replacement by slag, %	W/(C + S)	Setting time, h:min		Total bleeding water cm³/cm² × 10⁻²
					Initial	Final	
1 C	—	—	0	0.50	4:56	6:27	4.16
2 ST	Pelletized (Canada)	4200	25	0.50	4:49	7:15	4.29
3 AT	Granulated (USA)	5400	25	0.50	4:23	6:43	3.08
4 AL	Granulated (Canada)**	3700	25	0.50	4:54	7:52	6.11
5 AL	Granulated (Canada)**	4600	25	0.50	4:49	7:20	5.67
6 AL	Granulated (Canada)**	6080	25	0.50	5:30	8:08	5.00
7 ST	Pelletized (Canada)	4200	50	0.50	5:54	9:32	4.13
8 AT	Granulated (USA)	5400	50	0.50	5:13	8:00	3.33
9 AL	Granulated (Canada)**	3700	50	0.50	6:06	9:39	9.50
10 AL	Granulated (Canada)**	4600	50	0.50	—	—	6.20
11 AL	Granulated (Canada)**	6080	50	0.50	6:16	9:37	2.78

*From reference (46).
**Same source but ground to different fineness
Note: Cement type: ASTM Type I, C.A.: crushed limestone 19–mm max size
F.A.: natural sand; A.E.A.: sulphonated hydrocarbon type.

PLASTIC-SHRINKAGE CRACKING
OF FRESH CONCRETE

Plastic-shrinkage cracking occurs in fresh concrete under conditions which cause a net removal of water from exposed concrete surfaces, thus creating tensile stresses beyond the low early-age tensile capacity of concrete. As concrete containing silica fume shows little or no bleeding, thus allowing very little water rising to the surface, the risk of cracking is high in fresh concrete. Shrinkage cracking of fresh concrete can be a very serious problem under curing conditions of elevated temperatures, low humidity, and high winds, allowing rapid evaporation of water from freshly placed concrete. Johansen (58) and Sellevold (59) have reported that fresh concrete mixtures containing silica fume are most vulnerable to plastic shrinkage cracking as they approach initial set. To overcome this problem, the surface of concrete should be protected from evaporation by covering with plastic sheets, wet burlap, curing compounds, and evaporation retarding admixtures.

Unlike silica fume concrete, the concretes incorporating fly ash, slag or natural pozzolans do not exhibit plastic-shrinkage cracking during their setting.

CHAPTER 6

Effect of Mineral Admixtures on Properties of Hardened Concrete

Effects of mineral admixtures on the following properties of hardened concrete are discussed in this chapter: Colour, strength (compressive, flexural, and bond), Young's modulus of elasticity, creep, drying shrinkage, and thermal shrinkage.

COLOUR

The incorporation of mineral admixtures in concrete affects its colour on hardening, and this depends on the type and amount of the admixture used. For example, the use of both fly ash and silica fume results in the hardened concrete being somewhat darker than the conventional portland cement concrete. This colour difference is more evident on the surface of the wet hardened concrete than in dry concrete. Also, in concrete incorporating high levels of either fly ash or silica fume with high carbon content, the dark colour is more pronounced. The rice-husk ash concrete is significantly darker than the portland cement concrete when there is high carbon content in the ash. On the contrary, concrete incorporating slag or natural pozzolan is generally lighter in colour than normal portland cement concrete. When slag concrete is broken in flexure or compression, the interior of the broken specimens exhibits deep bluish-green colour. After sufficient exposure to air, the colour disappears. The degree of colour, which results from the reaction of the sulphides in the slag with other compounds in cement, depends upon the percentage of slag used, curing conditions and the rate of oxidation.

COMPRESSIVE STRENGTH

The use of mineral admixtures affects significantly the strength properties of concrete, and some of the factors that determine the rate of strength development are discussed below:

Fly Ash

Many variables influence the strength development of fly-ash concrete, the most important being the following:

- type of fly ash (chemical composition),
- particle size,
- reactivity,
- temperature of curing.

Effect of Fly Ash Type on Concrete Strength

In general, the rate of strength development in concrete tends to be only marginally affected by high-calcium fly ashes. A number of authors have noted that concrete incorporating high-calcium fly ashes can be made on an equal weight or equal volume replacement basis without any significant effect on strength at early ages (60,61). Yuan and Cook (60) have examined the strength development of concretes with and without high-calcium fly ash (CaO = 30.3%). The data from their research are shown in Fig. 6.1.

Low-calcium fly ashes, ASTM Class F, were the first to be examined for use in concrete, and much that has been written

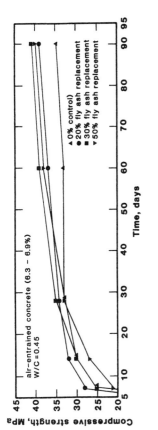
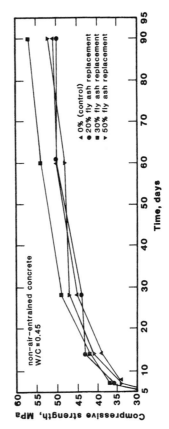

FIGURE 6.1 Compressive strength development of concretes containing high-calcium fly ash. From reference 60.

on the behaviour of fly ash concrete has been determined by using Class F ashes. In addition, the ashes used in much of the early work came from older power plants and were coarser in particle size, contained a higher content of unburned carbon, and were often relatively less active as pozzolans. When used in concrete by employing simple replacement methods, these ashes showed exceptionally slow rates of strength development. Such observations led to the development of generalized opinions that "fly ash reduces strength at all ages" (62). Conversely, they have resulted in considerable efforts to understand the factors affecting strength of fly ash concrete, and the ways in which fly ash ought to be proportioned in concrete mixtures in order to obtain desired rates of strength gain.

Particle size

Particle size can influence strength development in two ways. First, particles larger than 45 μm appear to influence water requirement in an adverse way. Thus, they act counter to the needs of the proportioning methods used to compensate for the slow rate of reaction of fly ash at early ages. Secondly, cementing activity occurs on the surface of the solid phases, through processes involving the diffusion and dissolution of materials in concentrated pastes. Surface area must play a considerable role in determining the kinetics of such processes. Mehta (10) found that the compressive strength of mortars containing fly ash was directly proportional to the percentage of < 10 μm particles in the fly ash.

Reactivity

Pozzolanic activity, although well established as a phenomenon, is far from being well understood. Much of the research currently in progress on fly ash is directed to understanding pozzolanic reactions. It is not within the scope of this report to discuss this work in any detail, and the reader is referred to

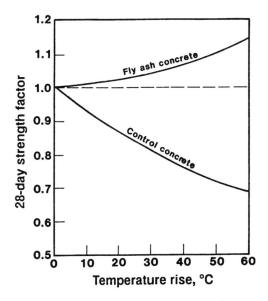

FIGURE 6.2 Effect of temperature rise during curing on the compressive strength development of concretes. From reference 67.

the other published sources (2,63,64,65) for material on this subject.

Temperature of Curing

When portland cement concrete is cured at temperatures in excess of 30°C, an increase is seen in strength at early ages but a marked decrease in strength is noted in the mature concrete (66). Concrete containing fly ash behaves significantly differently. Fig. 6.2 shows the general way in which the temperature, reached during early ages of curing, influences the 28–day strength of concrete. In contrast to the loss of strength that occurs with ordinary portland cement, fly ash concretes

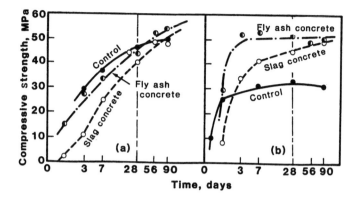

FIGURE 6.3   Compressive strength development of concretes: (a)
cured under normal conditions; (b) cured by temperature matching. From reference 67.

show strength gain as a consequence of heating. This is of
great value in the construction of mass concrete or in concrete
construction at elevated temperatures.

Bamforth (67) has reported in-situ and laboratory-observed
effects of temperature on the strength development of mass
concrete containing fly ash or slag substitutes. Fig. 6.3(a)
shows the strength development of standard cubes made from
the three concretes under standard laboratory curing conditions. Approximate strength equality was reached at 28 days
and, in fact, the strength of the fly ash concrete was close to
equality with the control concrete at earlier ages.

Fig. 6.3(b) shows the effects of curing under temperature-
matched conditions, where the effects of the early-age temperature cycle at the centre of the concrete mass were simulated.
In all cases, early strength development was accelerated. However, at 28 days, whereas the strength of the fly ash concrete
was enhanced by temperature, the strength of the control con-

crete was significantly less, being 30% below that of the standard water-cured concrete.

Silica Fume

It is well recognized that silica fume can contribute significantly to the compressive strength development of concrete. This is because of the filler effect and excellent pozzolanic properties of the material which translate into a stronger transition zone at the paste-aggregate interface. The extent to which silica fume contributes to the development of compressive strength depends on various factors such as percentage of silica fume, water-to-cementitious materials ratio W/(C + SF), cementitious materials content, cement composition, type and dosage of superplasticizer, temperature, humidity, and age of curing.

Superplasticizing admixtures play an important role in ensuring an optimum strength development of silica fume concrete. The water demand of silica fume concrete is directly proportional to the amount of silica fume used if the slump of concrete is to be maintained constant by increasing the water content rather than by using a superplasticizer. In such instances, the increase in the strength of silica fume concrete over that of control concrete is largely offset by the higher water demand, especially for high silica fume content at early ages, as illustrated in Fig. 6.4. In general, the use of a superplasticizer is a prerequisite in order to achieve proper dispersion of the silica fume in concrete, and to fully utilize its contribution to the strength potential. In fact, many important applications of silica fume in concrete depend strictly upon its utilization in conjunction with superplasticizing admixtures.

Silica fume concretes have compressive strength development patterns which are generally different from those of portland cement concrete. The strength development characteristics of these concretes are somewhat similar to those of fly ash concrete, except that the effects of the pozzolanic

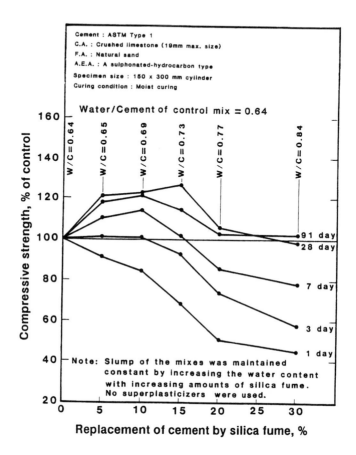

FIGURE 6.4   Relation between compressive strength of concrete
and dosage of silica fume, (W/C = 0.64). From reference 42.

reactions of the former are evident at earlier ages. This is due
to the fact that silica fume is a very fine material with a very
high amorphous silica content. The main contribution of sili-

ca fume to concrete strength development at normal temperatures takes place between the ages of about 3 and 28 days. The overall strength development patterns can vary according to concrete proportions and composition, and are also affected by the curing conditions.

Carette and Malhotra (68) have reported investigations dealing with the short- and long-term strength development of silica fume concrete under conditions of both continuous water-curing and dry-curing after an initial moist-curing period of 7 days. Their investigations covered superplasticized concretes incorporating 10% silica fume as a replacement by mass for portland cement, and W/(C + SF) ranging between 0.25 and 0.40. The compressive strength development at ages up to 3.5 years under continuously wet curing conditions is illustrated in Fig. 6.5, for concrete with a W/(C + SF) of 0.40. As expected, the major contributions of silica fume to the strength take place prior to 28 days with the largest gains in strength of the silica fume concrete over the control concrete being recorded between the ages of 28 and 91 days. However, this gain progressively diminishes with age. For concretes with a W/(C + SF) of 0.30 and 0.40, it largely disappeared at later ages. Under air-drying conditions, the strength development pattern was found not to be significantly different from that of water-cured concretes up to the age of about 91 days; thereafter, however, air-drying clearly had adverse effect on the strength development of both types of concrete. The effect was generally more severe for silica fume concrete where some reduction in strength was recorded between the ages of 91–day and 3.5 years, especially for concrete with a W/(C + SF) of 0.30 and 0.40. These trends of strength reduction have not been clearly explained yet, but they appear to stabilize at later ages, and therefore may be of little practical significance.

Curing temperatures have also been shown to affect significantly the strength development of silica fume concrete. This aspect has been examined in some detail by several investigators. In general, these investigations have indicated that the

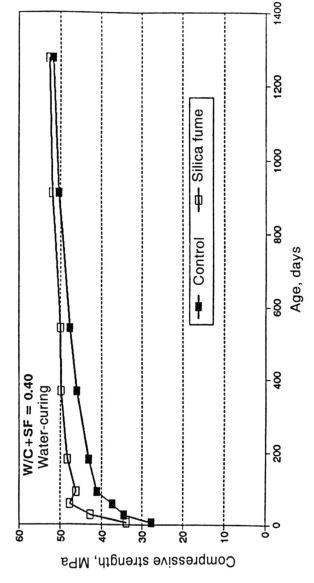

FIGURE 6.5   Compressive strength development of water-cured concrete – W/C + SF = 0.40. From reference 68.

pozzolanic reaction of silica fume is very sensitive to temperature, and elevated temperature curing has a greater strength accelerating effect on the silica fume concrete than on comparable portland cement concrete (69,70).

The dosage of silica fume is obviously an important parameter influencing the compressive strength of silica fume concrete. For general construction, the optimum dosage generally varies between 7 and 10%; however, in specialized situations up to 15% silica fume has been incorporated successfully in concrete.

Silica fume together with fly ash and high dosages of superplasticizers have been incorporated into concrete to produce compressive strength exceeding 100 MPa (71). The proper selection of cement, aggregates and the chemical admixtures plays an important role in developing mixture proportions for this type of concrete. The mixture proportioning data by Bürg and Ost (72) are shown in Table 6.1, and the compressive strengths are shown in Table 6.2.

Slag

The compressive strength development of slag concrete depends primarily upon the type, fineness, activity index, and the proportions of the slag used in concrete mixtures. Other factors that affect the performance of slag in concrete are the water-to-cementitious materials ratio and the type of cement used. In general, the strength development of concrete incorporating slags is slow between one to seven days compared with that of the control concrete. Between 7 and 28 days, the strength approaches that of the control concrete; beyond this period, the strength of slag concrete exceeds the strength of control concrete. Fig. 6.6 shows compressive strength development with age for the granulated slag concrete for water-to-(cement + slag) ratio of 0.55. It is noted that the highest strength gain at 28 days was for concrete with the slag content at 40% by mass as replacement for cement.

TABLE 6.1   Concrete Mixtures*

Parameter, units per cubic meter	Mix number					
	1	2	3	4	5	6
Cement fume, kg (2)	564	475	487	564	475	327
Silica fume, kg (2)	—	24	47	89	74	27
Fly ash, kg.	—	59	—	—	104	87
Coarse agg. SSD, kg (3)	1068	1068	1068	1068	1068	1121
Fine agg. SSD, kg	647	659	676	593	593	742
HRWR Type F, litre	11.60	11.60	11.22	20.11	16.44	6.30
HRWR Type G, litre	—	—	—	—	—	3.24
Retarder Type D, litre	1.12	1.05	0.97	1.46	1.50	—
Total water, kg (4)	158	160	155	144	151	141
Water:cement ratio	0.28	0.34	0.32	0.25	0.32	0.43
Water:Cementitious ratio	0.28	0.29	0.29	0.22	0.23	0.32

(1)   As reported by a ready–mixed concrete supplier
(2)   Dry weight
(3)   Maximum aggregate size: Mixtures 1–5, 12 mm; Mixture 6, 25 mm
(4)   Weight of total water in the mixtures including admixtures
*From reference 72

Malhotra et al (73) have reported an investigation in which
small amounts of silica fume have been added to pelletized
slag concrete to increase the early-age strength. Fig. 6.7 illus-
trates the strength development of concrete from 1 to 180 days.
The authors concluded:

- The low early-age strength of portland cement concrete
  incorporating blast-furnace slag can be increased by the
  incorporation of condensed silica fume. The gain in
  strength is, in general, directly proportional to the percentage of the fume used.

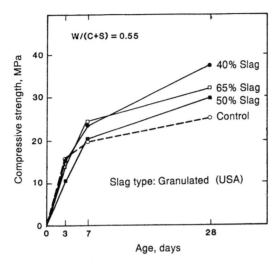

FIGURE 6.6   Age versus compressive strength relationship for air-
entrained concrete – W/(C + BFS) = 0.55. From Reference 57.

- At three days, the increase in strength is generally mar-
  ginal, especially for concrete with a high W/(C + BFS).
  However, at the age of 14 days and beyond, with minor
  exceptions, the loss in compressive strength of concrete
  due to the incorporation of BFS can be fully compensated for with a given percentage of condensed silica
  fume, regardless of the W/(C + BFS). This is also true
  for the flexural strength.

- The continuing increase in strength at 56, 91 and 180
  days of the concrete incorporating BFS and condensed
  silica fume indicates that sufficient calcium hydroxide
  (liberated during the hydration of portland cement) is
  present at these ages for the cementitious reaction to continue.

TABLE 6.2   Compressive Strength Development[*]

| Age | Compressive strength of moist cured 152 × 305–mm specimens, MPa | | | | | |
| | Mix number | | | | | |
days	1	2	3	4	5	6
3	58	58	55	75	56	42
7	66	72	73	96	79	60
28	79	89	92	119	107	73
56	81	97	94	121	112	n.c.s.[*]
91	87	100	96	132	119	89
182	92	104	95	134	124	n.s.c
272	98	108	98	135	130	n.s.c.
426	105	117	99	130	123	n.s.c
578	106	115	101	131	121	n.s.c
1085	110	116	110	125	126	n.s.c

| Age | Compressive strength of air cured 152 × 305–mm specimens, MPa | | | | | |
| | Mix number | | | | | |
days	1	2	3	4	5	6
56	90	100	98	119	117	n.s.c.
91	90	105	99	131	120	n.s.c.
182	95	106	100	i.d.[†]	i.d.	n.s.c.
272	94	106	99	136	121	n.s.c.
426	94	102	96	124	118	n.s.c.
1085	94	110	95	124	116	n.s.c.

Natural Pozzolans

Although the pozzolanic reactions in a portland-pozzolan cement paste begin as soon as alkalies and calcium ions are released during the hydration of portland cement, most of the pozzolanic activity, and therefore the strength development associated with it, seems to occur after seven days of hydra-

TABLE 6.2   (Continued)

Compressive strength of moist cured 152 × 305–mm specimens, MPa					
Age days	Mix number					

Age days	1	2	3	4	5	6
3	57	54	55	72	53	43
7	67	71	71	92	77	63
28	79	92	90	117	100	85
56	84	94	95	122	116	n.s.c.
91	88	105	96	124	120	92
182	97	105	97	128	120	n.s.c.
272	96	110	106	132	131	n.s.c.
426	103	118	100	133	119	n.s.c.
576	108	116	108	144	132	n.s.c.
1085	115	122	115	150	132	n.s.c.

Compressive strength of air cured 152 × 305–mm specimens, MPa					
Age days	Mix number					

Age days	1	2	3	4	5	6
56	90	97	104	121	111	n.s.c.
91	95	108	105	132	119	n.s.c.
182	97	102	99	i.d.	i.d.	n.s.c.
272	86	93	101	139	125	n.s.c.
426	93	104	99	134	126	n.s.c.
1085	95	104	95	136	119	n.s.c.

*From Reference 72

tion (74,75). Typical results from an investigation on the effect of curing time on the compressive strength of ASTM C 109 mortars made with portland-pozzolan cements containing 10, 20, or 30% Santorin earth, are presented in Fig. 6.8 and 6.9 (74).

From the compressive strength data for 1,3,7, and 28 days (Fig. 6.8), it is evident that up to seven days, the strength was

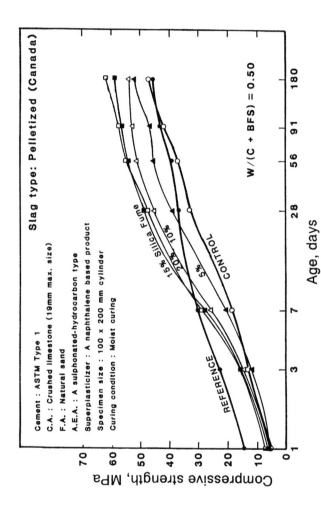

FIGURE 6.7 Age versus compressive strength relationship for concrete incorporating silica fume and pelletized slag – W/(C + BFS) = 0.50. From reference 73.

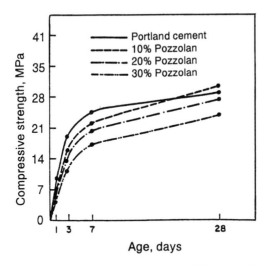

FIGURE 6.8   Compressive strengths up to 28 days of cements made with Santorin earth. From reference 74.

almost proportional to the amount of portland cement present in the blended cement. This result shows that in seven days of hydration, the pozzolanic reaction had not progressed far enough to influence the strength. At 28 days, however, the strength of the cement containing 10% pozzolan was slightly higher than the reference portland cement; the cements containing 20 and 30% pozzolan showed 7 and 18% lower strengths than the control, respectively. It was concluded from the data in Fig. 6.8 that during the 7- to 28-day hydration period, the beneficial effect on strength resulting from the pozzolanic reaction was not considerable.

The data for 28-, 90-, and 365 days in Fig. 6.9 shows that at 90 days, the strengths of cement/mortars with both 10 and 20% Santorin earth were about 10% greater than the reference portland cement. The one-year strength of the 30% pozzolan cement/mortar was similar to the reference portland cement

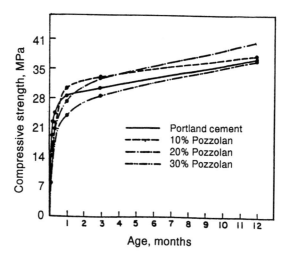

FIGURE 6.9 Compressive strengths up to 12 months of cement/
mortars made with Santorin earth. From reference 74.

whereas the other two pozzolan cements showed higher
strengths than the reference cement. In Fig. 6.10, similar results on the effect of substituting portland cement with an Italian natural pozzolan were reported (75). The authors
confirmed the slow reaction between the lime and pozzolan
and concluded that at early ages, the cements containing pozzolans, show lower strengths than the reference portland cement; however, the ultimate strength can be higher depending
on the quality and quantity of the pozzolan used. It seems that
an excess of pozzolan, i.e., more than 30% by mass in the
blended cement, should be avoided when a substantial reduction in the mechanical strength of the product cannot be toler-
ated, especially at early ages and under cold weather
conditions.

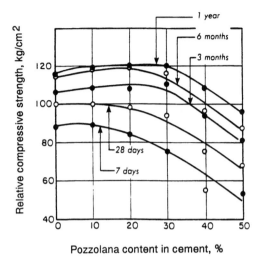

FIGURE 6.10   Effect of substituting an Italian natural pozzolan for portland cement on the compressive strength of ISO mortar. From reference 75.

Rice-Husk Ash

The compressive strength development data* for the rice-husk concrete are shown in Table 6.3. The rice-husk ash concrete has marginally higher compressive strengths at various ages up to 180 days compared with that of the control concrete. For example, the rice-husk ash concrete incorporating 10% ash as replacement for portland cement had compressive strengths of 38.6 and 48.2 MPa at 28 and 180 days, respectively; the corresponding strengths for the control concrete were 36.4 and 44.2 MPa, respectively (Table 6.3).

---

*CANMET Division Report MSL 95–007 (OP&J)

TABLE 6.3   Mechanical Properties of Rice-Husk Ash Hardened Concrete*

Mix No.	RHA content (%)	Silica fume content (%)	W/C or W/C+RHA	Unit weight (kg/m³)	Strength properties (MPa)								E modulus (GPa)
					Compressive						Splitting-tensile	Flexural	
					1d	3d	7d	28d	90d	180d	28d	28d	28d
C0	0	0	0.40	2350	20.9	25.5	28.9	36.4	42.5	44.2	2.7	6.3	29.6
R10	10	—	0.40	2320	22.1	26.2	31.1	38.6	47.0	48.3	3.5	6.8	29.6

*From CANMET unpublished data
Compressive strength – average of three 102×203–mm cylinders
Splitting–tensile strength – average of two 152×305–mm cylinders
Flexural strength – average of two $102 \times 76 \times 406$–mm prisms
"E" modulus – average of two 152×305–mm cylinders

FLEXURAL STRENGTH

Fly Ash and Silica Fume

In general, the available data indicate that the development pattern of flexural strength of fly ash or silica fume concrete under normal curing conditions is essentially the same as that for portland cement concrete (76 to 79). Based on the Norwegian experience, Gjørv (80) concluded that the ratio of flexural-to-compressive strength was similar for concretes with or without silica fume at least within the range of structural applications. It was, however, pointed out that early drying affects more the flexural strength of silica fume concrete than that of the portland cement concrete. Other investigators have reported similar findings (68,81,82). Some data on the flexural strength of rice-husk ash concrete are shown in Table 6.3.

Slag

In general, at seven days and beyond, the flexural strength of concrete incorporating slag is comparable to or greater than the corresponding strength of control concrete (57). The increased flexural strength of the slag concrete is in part due to the stronger bonds in the slag-cement/aggregate systems because of the shape and surface texture of the slag particles.

BOND STRENGTH

Fly Ash

There is not much published data on the bond strength of concrete containing fly ash. Those fly ashes which reduce bleeding of concrete will, no doubt increase the bond of concrete to reinforcement; on the other hand, coarser fly ash which may cause more bleeding in concrete is likely to do the opposite.

The ACI Committee 232 report (83) on fly ash discusses the bond of fly ash concrete to steel as follows:

> "The bond of adhesion of concrete to steel is dependent on the surface area of the steel in contact with the concrete, the location of reinforcement, and the density of the concrete. Fly ash usually will increase paste volume and reduce bleeding. Thus, the contact at the lower interface where bleed water typically collects may be increased, resulting in improved bond. Development length of reinforcement in concrete is primarily a function of concrete strength. With proper consolidation and equivalent strength, the development length of reinforcement in concrete with fly ash should be at least equal to that in concrete without fly ash. These conclusions about bond of concrete to steel are based on extrapolation of what is known about concrete without fly ash. The bonding of new concrete to old is little affected by the use of fly ash."

Silica Fume

The bond-strength characteristics of silica fume concrete have been studied by a number of researchers, and there is a clear agreement that silica fume significantly improves the bond strength at the paste-aggregate as well as at the paste-steel reinforcement interfaces. In their study on the bond effects on high-strength silica fume concretes, Bentur and Goldman (84) concluded that silica fume has an inherent effect on strength enhancement, which increases with an increase in silica fume content. The effect was attributed to the improved aggregate-matrix bond resulting from the formation of a less porous transition zone in the silica fume concrete. The denser transition zone at the paste-aggregate interface in silica fume concrete has also been observed by several other investigators (85–90). Similarly, Gjørv et al. (91) reported that the presence of silica fume affects the morphology and microstructure of the steel-cement paste transition zone which was found to be reduced in thickness and had a lower porosity. The authors suggested that the stronger steel-concrete bond in silica fume concrete was due to the improvement in the transition zone. The enhanced bond characteristics at the paste-reinforcement interface have also been reported for lightweight silica-fume concretes (92,93). Silica fume is reported to improve the bond between cement paste and various types of reinforcing fibres (69,94,95,96).

Slag and Natural Pozzolans

There is little information available on the bond of steel reinforcement to concrete incorporating slags or natural pozzolans. Once again, the bond characteristics will generally be related to the ultimate strength properties of the concrete.

YOUNG'S MODULUS OF ELASTICITY

Based on the published data by various investigations, there appears to be no significant difference between Young's modulus of elasticity of concrete with or without mineral admixtures at 28 days. However, like compressive strength, concrete incorporating fly ash or slag has lower modulus at early strength, and higher modulus at ultimate strength when compared with the control concrete. Fig. 6.11 shows typical stress-strain relationship for concrete both with and without fly ash (51).

Table 6.4 shows modulus of elasticity data by Wainright and Tolloczko (97) on concrete incorporating a slag from the U.K. At early ages, and at higher percentages of slag, the concrete shows lower modulus in relation to the control concrete.

Limited data on the "E" modulus of the rice-husk ash concrete are shown in Table 6.3; the values for the control and the rice-husk concretes are comparable.

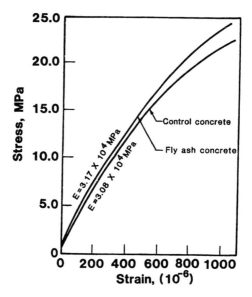

FIGURE 6.11   Stress-strain relationship for concretes. From reference 51.

TABLE 6.4   Static modulus of elasticity*

Slag content					
0%		50%		70%	
Age, days	E, GPa	Age, days	E, GPa	Age, days	E, GPa
1	26.5	1	17.0	1	6.5
2	31.5	2	20.0	2	14.0
4	34.0	3	25.0	3	18.5
7	36.0	7	28.5	7	25.5
31	38.5	28	37.0	28	32.0
56	39.5	56	39.0	56	35.0
171	41.0	165	42.0	174	39.5

*From reference (97).
Note: Concretes were proportioned to have minimum cube compressive strength of 45 MPa at 28 days.

CREEP

The published data on the creep behaviour of concrete incorporating mineral admixtures are sparse.

Ghosh and Timusk (98) have examined ASTM Class F fly ashes of different carbon content and fineness values in concrete mixtures at nominal strength levels of 20, 35, and 55* MPa (water-to-cement ratios 1.0, 0.4, and 0.2, respectively). Each concrete was proportioned for equivalent strength at 28 days. In the majority of specimens, it was found that fly ash concretes showed less creep than the reference concretes. This was attributed to a relatively higher rate of strength gain after the time of loading than for the reference concretes.

The published data on the creep strain of silica fume concrete are limited. Bilodeau, Carette and Malhotra (99) reported an investigation on the mechanical properties, creep, and drying shrinkage of high-strength concretes incorporating fly ash, slag, and silica fume. The cementitious materials content of the concretes was about 530 kg/m^3, and the water-to-cementitious materials ratio was 0.22. In addition to the reference concrete, concrete mixtures with silica fume at 7% and 12% cement replacements, fly ash at 25% cement replacement, slag at 40% cement replacement, combination of 7% silica fume and 25% fly ash, and 7% silica fume and 40% slag were investigated. The test specimens were subjected to creep loading after 35 days of moist curing at an applied stress equivalent to 35% of the compressive strength. The results obtained in this investigation are shown in Table 6.5. After one year duration, the creep strains of the reference, 7% silica fume, and 12% silica fume concretes were 1505×10^{-6}, 713×10^{-6}, and 836×10^{-6}, respectively. For concrete with 7% silica fume and 40% slag, the creep strain measured was the lowest at 641×10^{-6}. The

---

*Cement of low heat of hydration was used.

TABLE 6.5   Summary of Creep Test Results[*]

Mixture No.	W (C+SCM)	Silica Fume, %	Fly Ash, %	Slag, %	Age at Loading, Days	Applied Stress, MPa	Duration of Loading, Days	Initial Elastic Strain, $\times 10^{-6}$	Creep[+] Strain, $\times 10^{-6}$
1	0.22	0	0	0	35	28.0	385	717	1505
2	0.22	7	0	0	35	28.0	380	677	713
3	0.22	12	0	0	35	28.0	378	694	836
4	0.22	0	25	0	36	28.0	372	750	1069
5	0.22	7	25	0	35	28.0	371	645	749
6	0.22	0	0	40	35	28.0	366	690	1005
7	0.22	7	0	40	35	28.0	364	632	641

+Creep strain = total load induced strain − initial elastic strain
*From reference 99

pore structure and low quantity of free water in these concretes could have resulted in very low creep values, i.e. about 40% of the creep of the reference concrete.

Bamforth (67) has reported limited data on creep strains of concrete mixtures with and without fly ash and granulated slags, loaded to a constant stress-strength ratio of 0.25 (Fig. 6.12). He found that for concretes loaded at an age greater than 24 h, the effects of fly ash and slag were to reduce significantly the magnitude of the creep.

DRYING SHRINKAGE

The drying shrinkage of portland cement concrete incorporating a mineral admixture is dependent on the hydration products and water demand of the mixtures. Laboratory investigations should be performed to determine the drying shrinkage of test specimens made with the portland cement and the type of mineral admixture to be used on a particular project. Some of the published information is summarized below.

Fly Ash

It has been generally reported that the use of fly ash in normal proportions does not significantly influence the drying shrinkage of concrete. Ghosh and Timusk (98) showed that for the same maximum size of aggregate and for all strength levels, the shrinkage of concrete containing fly ash is lower than that for non-fly ash concrete.

In their studies of concrete using high-calcium fly ash, Yuan and Cook (100) concluded that the replacement of cement by fly ash had little influence on drying shrinkage. Their data are shown in Fig. 6.13. Similar conclusions may also be drawn from the data on shrinkage obtained by Carette and Malhotra

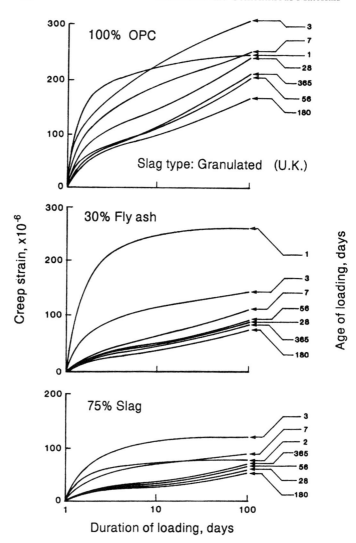

FIGURE 6.12 Creep of concrete with and without granulated
blast-furnace slag loaded to constant stress-strength
ratio of 25%. From reference 67.

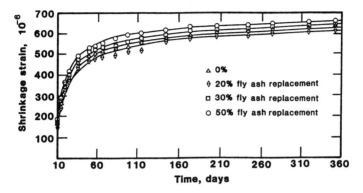

FIGURE 6.13   Drying shrinkage of concretes incorporating high-calcium fly ash. From reference 60.

(32) for concrete mixtures incorporating a variety of fly ashes (Table 6.6).

Silica Fume

Investigations carried out on the drying shrinkage of silica-fume concretes indicate that the long-term drying shrinkage of the concretes is not significantly affected by the silica fume addition. In an investigation carried by Johansen (77), concretes with silica fume content up to 25% and W/(C + SF) ranging from 0.37 to 1.06 were exposed to a drying environment (50% R.H.) at the ages of 1 and 28 days. It was found that when the W/(C + SF) was lower than 0.60, the drying shrinkage was about the same for the reference concrete and the silica-fume concrete. Some of these mixtures incorporated water reducers. However, the shrinkage values were found to be higher for concretes containing 25% silica fume and no water reducer. In another investigation, Pistilli et al (56) monitored the drying shrinkage of silica fume concretes. At 64 weeks, the concrete with 237 kg/m³ cement and a W/(C

TABLE 6.6   Drying Shrinkage of Fly Ash Concrete[*]

Mixture No.[**]	Duration of drying, days	Initially cured for 7 days in water		Initially cured for 91 days in water	
		Moisture[***] loss, %	Drying shrinkage, $\times 10^{-6}$	Moisture[***] loss, %	Drying shrinkage, $\times 10^{-6}$
Control 2	224	55.0	422	53.7	453
F1	224	57.5	447	47.9	365
F2	224	57.3	364	45.4	280
F3	224	56.9	411	56.2	405
F4	224	54.7	379	49.2	387
F5	224	58.8	404	51.1	403
F6	224	60.6	475	56.4	454
F7	224	64.3	397	54.1	433
F8	224	56.3	400	—	327
F9	224	58.2	390	49.3	361
F10	224	58.4	642	55.2	500
F11	224	49.5	454	48.9	362

[*]From reference 32.
[**]See Tables 5.1 for fly ash and mixture proportioning data.
[***]As a percentage of total original water.

+ SF) of 0.70 showed only a slight increase in the drying shrinkage when 24 kg/m³ of silica fume was incorporated, whereas for a concrete with 297 kg/m³ cement and 0.60 W/(C + SF), the drying shrinkage remained the same with or without the silica fume.

The initial curing period, and the conditioning of the test specimens before exposure to drying environment plays a critical part in the drying shrinkage of silica-fume concrete. Wolsiefer (78) monitored the drying shrinkage strains of high-strength concretes incorporating silica fume and high dosages of a superplasticizer. The concretes were moist cured for 1 and 14 days before the drying shrinkage measurements were made according to ASTM C 157. The drying shrinkage strains were found to be higher for specimens moist cured for one day. Maage (47) also reported that concretes containing silica fume exhibit higher drying shrinkage than the control concretes when these concretes were exposed to drying immediately after demoulding.

Fig. 6.14 shows the results from an investigation by Carette and Malhotra (48) on the drying shrinkage strain results of concretes made with a W/(C + SF) of 0.40 and incorporating 15, and 30% silica fume. The concretes were subjected to 28-day initial moist curing before the drying shrinkage tests (ASTM C 157). After 420 days of drying, the control concrete and the concrete with 15% silica fume showed about the same drying shrinkage strain, while the concrete with 30% silica fume showed slightly lower values.

The drying shrinkage strains of a normal-strength concrete (W/C of 0.57) and high-strength silica-fume concretes W/(C + SF) of 0.22, 0.25, 0.28) containing 10% silica fume were monitored by Tachibana et al. (101). At one year, the drying shrinkage strains of the high-strength concretes were in the range of 540 to 610×10^{-6}, whereas for the normal strength concrete the value was 930×10^{-6}.

In an investigation reported by Bilodeau et al (99), drying shrinkage tests were made on high-strength concretes after 7 days of moist curing. The concrete mixtures incorporated 7, and 12% silica fume, as well as combinations of silica fume, fly ash, and slag. Concretes containing silica fume, even in the presence of other mineral admixtures, showed lower drying shrinkage strains. The drying shrinkage strains of concrete incorporating 7% silica fume were found to be the lowest of all. de Larrard (102) attributes the phenomenon of low drying shrinkage in high-strength concretes to less free water in the system due to the low water-to-cementitious materials ratio. Such concrete will lose most of its free water primarily through hydration, resulting in a higher autogenous shrinkage and low drying shrinkage. As silica fume is incorporated in high strength concrete, the pores in the hardened concrete are re-

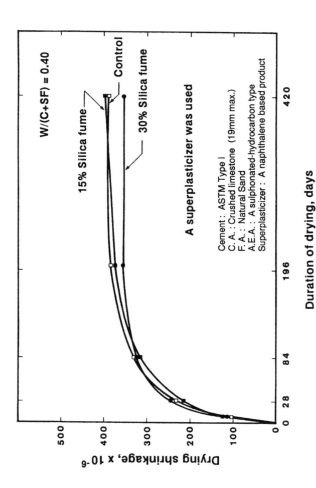

FIGURE 6.14   Relation between shrinkage and duration of drying for concrete—W/(C + SF)=0.40. From reference 42.

duced in size, thus increasing the surface tension, which results in increased autogenous shrinkage. Also, as the silica fume refines the pore size distribution of the hardened concrete, the permeability of the water vapour through the hardened paste is lowered, thus lowering the depth of drying at a given age (102).

Blast-Furnace Slag

Hogan and Meusel (57) have shown that the drying shrinkage of concrete incorporating granulated slag is more than that of the control concrete (Fig. 6.15 and 6.16). The increase in shrinkage is attributed to increased paste volume in the concrete when the slag which has lower specific gravity is used as replacement for portland cement on an equal mass basis. Fulton (103) has suggested that the shrinkage of concrete incorporating granulated slag can be reduced by taking advantage of the improved workability to increase the aggregate-to-cement ratio, or reduce the water-to-cement ratio of concrete.

Natural Pozzolans

In a study using concrete prisms and an 80-week drying period (74,104), it was found that compared to the reference concrete the drying shrinkage of the concrete made with blended cement incorporating 10, 20, or 30 per cent Santorin earth was not significantly different (Fig. 6.17).

Rice-Husk Ash

Limited available data* show that the drying shrinkage strain of the rice-husk ash concrete incorporating 10% ash as the replacement for cement after seven days of initial curing in

---

*CANMET Division Report MSL 95–007 (OP&J)

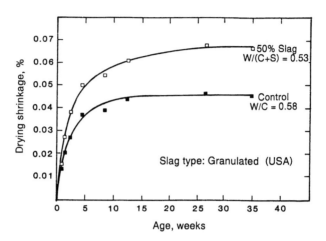

FIGURE 6.15   Drying   shrinkage   of   air-entrained   concrete—
W/(C + S) = 0.53. From reference 57.

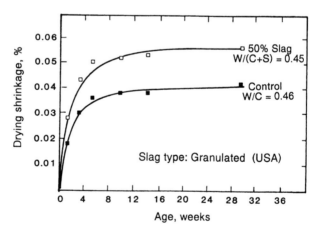

FIGURE 6.16   Drying   shrinkage   of   air-entrained   concrete—
W/(C + S) = 0.45. From reference 57.

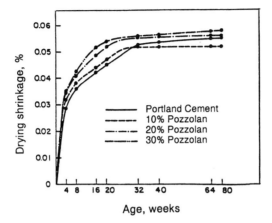

FIGURE 6.17   Drying shrinkage of concrete prisms made with cements containing various amounts of Santorin earth.
From reference 74.

lime-saturated water is of the order of 550×10^{-6} after 112
days, and is similar to the strains for the control concrete.

TEMPERATURE RISE AND THERMAL SHRINKAGE

The hydration or setting of portland cement paste is accompanied by evolution of heat that causes temperature rise in concrete. In general, with the exception of silica fume and
high-calcium fly ash, the replacement of portland cement by
mineral admixtures results in a significant reduction in the
temperature rise of both fresh and hardened concrete. This is
of particular importance in mass concrete where cooling, sub-
sequent to a large temperature rise, can lead to cracking.
Some aspects of temperature rise in concrete incorporating
mineral admixtures are discussed below.

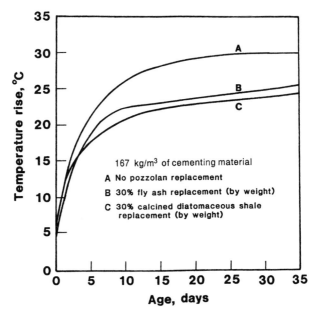

FIGURE 6.18   Influence of pozzolans on the temperature rise of
concrete. From reference 106.

Fly Ash

The first major use of fly ash in concrete was in the construction of a gravity dam where it was employed principally to
control temperature rise (105). Data reported by Elfert (106)
show the effects of a fly ash and a calcined diatomaceous
shale on the temperature rise of mass concrete (Fig. 6.18).
Compton and MacInnis (24) reported the temperature-time
curves shown in Fig. 6.19 for two experimental concretes,
one of which was made with 30% cement substitution by a
Canadian fly ash. Data on temperature rise of concrete con-

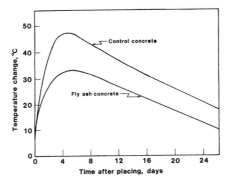

FIGURE 6.19   Temperature rise curves for fly ash and plain concrete test sections. From reference 24.

taining high volumes of fly ash has been published by Sivasundaram (107), and Sivasundaram et al (108).

Temperature rise, of course, depends upon more factors
than on the rate of heat generation alone; these include the rate
of heat loss and the thermal properties of the concrete and its
surroundings. Williams and Owens (109) have presented an es-
timation (Fig. 6.20) of the effect of the size of the elements on
the temperature rise in fly ash concretes.

Although it is reasonable to assume that all low-calcium fly
ashes will reduce the rate of temperature rise when used as cement replacement, high-calcium fly ashes do not necessarily
cause reduced heat evolution. Some high-calcium ashes, especially those containing significant amount of C_3A, alkali sulfates, or free CaO react very rapidly with water to generate
excessive heat. Crow and Dunstan (110) have reported the adiabatic temperature rise data on concrete mixtures (Fig. 6.21).
Whereas the concrete containing 25% of a low-calcium fly ash
showed a reduced rate of heat generation, concrete with 25% of
a high-calcium fly ash produced as much heat (at similar rate) as
the reference portland cement concrete. In general, the rate of
heat evolution parallels the rate of strength development.

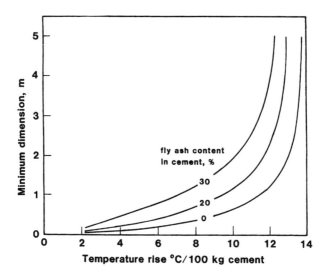

FIGURE 6.20   Effect of unit minimum size on the temperature rise
in fly ash and plain concrete. From reference 109.

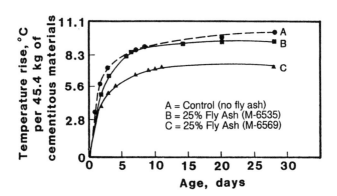

FIGURE 6.21   Adiabatic temperature rise in concrete made with
high-calcium fly ash. From reference 110.

Silica Fume

Researchers have shown that in concretes incorporating silica fume as cement replacement, a reduction in adiabatic heat is observed with no decrease in strength (111). A study by Tachibana et al. (101) confirmed that in high-strength concrete incorporating 540 kg/m^3 cement and 10% silica fume, the adiabatic heat rise was 9% lower than that of the concrete without silica fume. It has been shown however, that the addition of silica fume accelerates the temperature rise during the first 72 hours compared to a similar portland cement concrete (84). The overall temperature rise in later ages was found to be lower than that of the reference concrete. Such steep initial temperature rise in the silica fume concretes has been attributed to the accelerating effect silica fume has on the cement hydration.

Slag

Bamforth (67) has reported an extensive study of mass concrete containing fly ash or granulated blast-furnace slag as a substitute for cement. Included in this research was an in-situ investigation of the temperature rise and the resulting strain in three large foundation units.

Three concretes were examined.

- a control mixtures with a portland cement content of 400 kg/m^3;
- a mixture with 75% of the portland cement replaced by granulated slag;
- a mixture with 30% of the portland cement replaced by a bituminous fly ash.

The concretes were placed in three foundation units each 4.5 m deep, the volumes ranging from 144 to 212 m^3. The units were instrumented to measure temperature changes and movement during the early stages of the heat-generation cycle. The

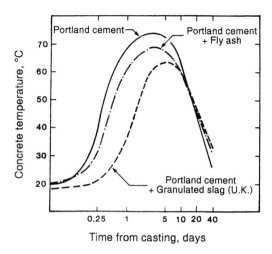

FIGURE 6.22   Variation of temperature recorded at mid-height in
              fly ash, slag, and plain concrete foundation units.

measured temperature variation in each unit at mid-height is
shown in Fig. 6.22.

Natural Pozzolans

Using a natural pozzolan from Italy, Massazza and Costa (22)
showed that the addition of the pozzolan to a portland cement
clearly reduced the heat of hydration; however, this reduction
was not directly proportional to the amount of cement replaced. It was smaller because of some evolution of heat during the pozzolanic reaction. For example, at the 20% level of
cement replacement, the 90-day heat of hydration of the portland cement was reduced from 94 to 85 cal/g, and the 28-day
heat of hydration was reduced from 85 to 76 cal/g. Nicolaidis
(39) found that the seven-day heat of hydration was similarly
reduced by 9 cal/g.

The ability of the pozzolanic materials to reduce the heat of hydration in portland-pozzolan cements is widely exploited in the construction of mass concrete structures, where the risk of thermal cracking is a major problem. For this purpose, as described earlier, natural pozzolans were used first in the United States from 1910 to 1912 for the construction of Los Angeles aqueduct, later in the construction of the San Francisco Bay and Golden Gate bridges, and the Friant, Bonneville, Davis, Glen Canyon, Flaming Gorge, Wanapum, and John Day dams (112).

Rice-Husk Ash

Fig. 6.23 shows the autogenous temperature rise with time in a rice-husk ash concrete mixture. The placing temperature of the concrete mixture was about 20°C. The maximum temper-

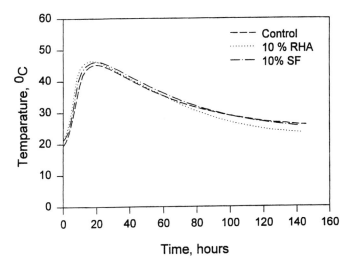

FIGURE 6.23   Autogenous temperature rise in 152 × 305-mm concrete cylinders.

ature for the concrete with the rice-husk ash (10% replacement for portland cement) was about 46°C which was almost the same as for the control concrete. The maximum temperature of the rice-husk ash concrete was reached after about 16 hr. which was earlier than that of the control concrete, indicating the high reactivity of the ash*.

---

*CANMET Division Report MSL 95–007 (OP&J)

CHAPTER 7

Concrete Incorporating Mineral Admixtures: Durability Aspects

Failure of concrete in a structure in a period less than its design life may be caused by external factors i.e. the environment to which it has been exposed or by a variety of internal causes. External factors may be physical or chemical in nature, such as weathering, extremes of temperature, abrasion, and exposure to aggressive chemicals. Internal causes may lie in the choice of materials or in inappropriate combinations of materials. Of all the causes of lack of durability in concrete, the most important is excessive permeability. Permeable concrete is vulnerable to attack by almost all classes of aggressive agents. To be durable, portland-cement concrete must be relatively impervious.

Increasingly, concrete is being selected for use as a construction material in aggressive or potentially aggressive environments such as sulphate and acidic waters. In modern times, the demands placed on concrete in marine environments have increased greatly, as concrete structures are being used in arctic, hot and arid, and tropical climates to contain and support the equipment, especially for oil and gas exploration and production. Concrete structures used to contain nuclear reactors must be capable of containing gases and vapours at elevated temperatures and pressures under emergency conditions. In all of these applications, mineral admixtures can play an important role in enhancing the durability of concrete. Therefore, a proper understanding of their effect on concrete durability is essential for their correct and economical application.

PERMEABILITY

Permeability of concrete is one of the most critical parameters in the determination of concrete durability to aggressive waters. As the permeability of concrete is lowered, its resistance to chemical attack increases. In a broader sense, permeability encompasses the transportation of liquid as well as gaseous phases into the interior of concrete, and it is evaluated by measuring any of the following: weight loss of saturated concrete over a period of drying, water transport under a pressure gradient, and rate of oxygen or chloride-ion diffusion.

As discussed in Chapter 4, the incorporation of fly ash, silica fume, slag, and natural pozzolans in concrete results in smaller crystalline products and finer pores in the hydrated cement paste especially at the aggregate/paste interface, leading to a decrease in permeability. This decrease is much higher in concretes incorporating silica fume or rice-husk ash due to their high pozzolanicity. Data from several investigations on the influence of various mineral admixtures on the permeability of concrete are given below.

Fly Ash

A number of investigations have been made concerning the influence of fly ash on the relative permeability of concrete containing fly ash as a replacement for cement. Davis (113) examined the permeability of a concrete pipe incorporating fly ash substituted for cement in amounts of 30 to 60%. Permeability tests were made on 150 × 150-mm cylinders at the ages of 28 days and 6 months; the results of these tests are shown in Table 7.1.

TABLE 7.1   Relative permeability of concretes with and without fly ash*

Fly ash			Relative permeability	
Type	% by weight	W/(C + F) by weight	28 days	6 months
None	—	0.75	100	26
Chicago	30	0.70	220	5
	60	0.65	1410	2
Cleveland	30	0.70	320	5
	60	0.69	1880	7

*From reference 113.

It is clear from the data that the permeability of the concrete was directly related to the quantity of hydrated cementitious product at any given time. After 28 days of curing, at which time little pozzolanic activity would have occurred, the fly ash concretes were more permeable than the control concretes. At 6 months, this was reversed, considerable imperviousness had developed, presumably due to the pozzolanic reaction of fly ash.

Silica Fume

In investigations reported by Gjorv (114) and Markestad (115), water permeability tests were conducted on silica-fume concretes with cement contents ranging from 100 to 500 kg/m³. When 10% silica fume was incorporated in the concrete with 100 kg/m³ cement, the permeability decreased from 1.6×10^{-7} to 4.0×10^{-10} m/s. The permeability of concrete containing 100 kg/m³ cement and 20% silica fume was found to be equivalent to that of a concrete containing 250 kg/m³ portland cement without any silica fume. It was also found by Gjorv (114) that the water permeability of concretes containing cement contents of 400 to 500 kg/m³ was in the

range of 10^{-14} to 10^{-15} m/s, irrespective of the silica fume
addition. Thus, in concretes with high cement contents, the
effect of silica fume on permeability appears to be negligible.

Permeability to chloride ions is critical in reinforced concretes exposed to sea water and de-icing salts. The chloride
ions destroy the passivating iron oxide layer over the reinforcement, thus making it vulnerable to corrosion. Therefore, by reducing the mobility of chloride ions into concrete, a measure
of protection against corrosion is provided. The incorporation
of mineral admixtures in concrete, in general, refines the pore
structure of the paste, and reduces the penetration of chloride
ions into concrete. ASTM C 1202 "Test Method For Electric-
Indication of Concrete's Ability to Resist Chloride-Ion Penetration" is the most commonly used in North America to
quantify the chloride-ion penetration into concrete.

Plante and Bilodeau (116) reported that the addition of 8%
silica fume significantly reduced the penetration of chloride
ions into concrete. With increasing cementitious materials content and decreasing water-to-cementitious materials ratio, the
chloride-ion penetration was reduced further. At a W/(C + SF)
of 0.21 and 500 kg/m³ cement and 40 kg/m³ silica fume, the
chloride-ion penetration in ASTM C 1202 corresponded to 196
coulombs at 28 days compared to 1246 coulombs for the reference concrete (Fig. 7.1). This reduction is primarily due to the
refined pore structure and increased density of the matrix.

In an investigation by Cohen and Olek (117), measurements
for the penetration of chloride ions into concretes, incorporating various forms of silica fume such as slurry, uncompacted,
and densified fume, were made. The control concrete was a
plain portland-cement concrete with 371 kg/m³ of ASTM Type
I cement, and 10% of the cement was replaced by silica fume
in the silica fume concretes. Water-to-cementitious materials
ratio was maintained at 0.35. The 28-day compressive strength
of the control concrete was about 43 MPa, and that of the silica
fume concretes was about 70 MPa. The chloride-ion penetration values measured at 28 days (using the ASTM 1202 test

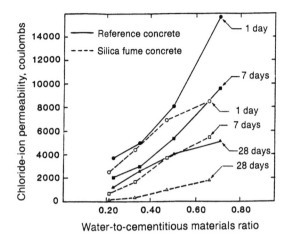

FIGURE 7.1   Influence of silica fume on chloride-ion permeability
of concrete. From reference 116.

method) were 1788 coulombs for the control concrete and 300
to 400 coulombs for the silica fume concretes. It is emphasized
that very low water-to-(cement + silica fume) ratios of the order of 0.30 and silica fume percentage of about 10% by weight
of cement are essential to obtain concretes which will provide
protection to reinforcing steel from corrosion. Berke (118) has
suggested that concretes with very low coulombs values alone
may not be immune to chloride-ion ingress, and thus corrosion
of embedded steel may occur.

Slags

The permeability of concrete depends mainly upon the
permeability of the cement paste which, in turn, depends
upon its pore-size distribution. Using mercury intrusion techniques, several investigators (119,120) have shown that the
incorporation of granulated slags in cement paste helps in the

transformation of large pores in the paste into smaller pores, thus resulting in decreased permeability of the matrix and, hence, of the concrete. Also, it is observed that granulated-slag concretes incorporating slags at up to 75% cement replacement have performed satisfactorily in exposure to sea water (121). Manmohan and Mehta (120) reported that hydrated cement pastes containing 70% slag were essentially impermeable.

Natural Pozzolans

Most work on the permeability of concrete containing finely divided mineral admixtures deals with natural pozzolans. Certain pozzolans are more effective than others in reducing permeability of concrete at early ages. However, under normal conditions of service, the permeability of concrete containing most natural pozzolans would be markedly reduced at later ages. Davis (122) concluded that in mass concrete, the use of a moderate to high proportion of a suitable pozzolan resulted in a degree of watertightness that was not otherwise obtainable. Part of the action of pozzolans in reducing permeability of concrete can be attributed to a reduction in water requirement, and to decreased segregation and bleeding. Mehta (9) performed permeability tests, and made pore-size distribution analysis with mercury intrusion porosimetry on 28–, 90–, and 365–day old pastes containing Santorin earth and portland cement. The data showing the effects of the age of hydration and amount of pozzolan present in the cement paste on a 3–hr water-penetration test are given in Table 7.2.

Rice-Husk Ash

In a recent investigation, Zhang and Malhotra* showed that portland cement concrete incorporating 10% rice-husk ash as

---

*CANMET Division Report MSL 95–007 (OP&J)

TABLE 7.2   Relative depth of penetration of water into hydrated cement pastes[*]

Hydration age	Depth of penetration, mm			
	Portland cement	10 percent Santorin earth	20 percent Santorin earth	30 percent Santorin earth
28 days	26	24	25	25
90 days	25	23	23	22
1 year	25	23	18	15

[*]From reference 9

replacement for cement had very high resistance to the penetration of chloride ions in the rapid chloride-ion permeability test (ASTM C 1202); the charge passed through the concrete was below 1000 coulombs. Mehta (7) reported less than 50 coulombs charge in rice-husk ash concrete specimens cured for one year.

CARBONATION

Calcium hydroxide, and to lesser extent, the calcium silicates and aluminates in hydrated portland cement react in moist conditions with carbon dioxide from the atmosphere to form calcium carbonate. The process, termed carbonation, reduces the alkalinity of concrete, eventually causing the destruction of the protective iron oxide film which is normally present at the surface of reinforcing steel. The rate at which concrete carbonates is determined by its permeability, degree of saturation with water, and the mass of calcium hydroxide available for reaction. Well-compacted and properly-cured concrete at a low water-to-cement ratio, will be sufficiently impermeable to resist the advance of carbonation beyond the first few millimeters from the surface of concrete.

In general, regardless of the type of mineral admixture being incorporated into concrete, the degree of carbonation of the concrete is no different from that of the control concrete with equivalent water-to-cement ratio and curing conditions. Notwithstanding the above, it should be mentioned that concrete incorporating mineral admixtures usually takes longer to reach the same level of maturity as normal portland cement concrete, and therefore proper curing of the former becomes more important.

It has been suggested that the use of mineral admixtures reduces the $Ca(OH)_2$ content of the concrete, and this promotes a faster rate of carbonation. However, this effect is offset by the very low permeability of the product, and this tends to impede the ingress of CO_2 in concrete. The net effect is usually favourable but it does depend upon various factors such as the amount and type of the mineral admixtures, water-to-cementitious materials ratio and curing conditions. All of these can have a significant influence on the $Ca(OH)_2$ content and permeability of concrete.

Ho and Lewis (123) examined the rates of carbonation of three types of concrete (plain, water-reduced, and fly ash) at equal slump. Accelerated carbonation was induced by storing specimens in an enriched 4% CO_2 atmosphere at 20°C and 50% RH for eight weeks. The authors assume that one week under these conditions is approximately equivalent to one year in a normal atmosphere (0.03% CO_2). The authors concluded as follows:

- Concretes having the same strength and water-to-cement ratio do not necessarily carbonate at the same rate.

- Based on a common 28-day strength, concrete containing fly ash showed a significant improvement in quality when curing was extended from 7 to 90 days. Such improvement was much greater than that achieved for the plain concrete.

- The depth of carbonation is a function of the cement content for concretes moist-cured for 7 days. However, with further curing to 90 days, concrete containing fly ash showed a slower rate of carbonation when compared to plain and water-reduced concretes.

Carette and Malhotra (68) compared the rate of carbonation of silica-fume concrete with that of reference portland cement concrete up to the age of 3.5 years. The test specimens had been initially moist-cured for 7 days before being stored under ambient drying conditions. At the water-to-cementitious materials ratio of 0.25, they found that all concretes remained free of any noticeable carbonation during the 3.5 year period, whereas at a W/(C + SF) of 0.40, all specimens exhibited signs of carbonation, the effect being slightly more marked for the silica- fume concrete. In general, however, it is agreed that carbonation is not a problem in adequately-cured, high-quality low W/C concrete, and this applies also to concrete-containing mineral admixtures.

SULPHATE RESISTANCE AND ACID RESISTANCE

The sulphate and acid resistance of concrete incorporating mineral admixtures is affected by the type and amount of cement, type and amount of mineral admixture, physical characteristics of the admixture, water-to-cementitious materials ratio, and curing conditions. In general, the use of mineral admixtures will increase the chemical resistance of concrete due mostly to the reaction of the admixture with calcium hydroxide to form additional calcium silicate hydrate (C–S–H) which fills in capillary pores in the cement paste, thus reducing the permeability of the system and the ingress of aggressive ions.

Fly Ash

In general, ASTM Class F fly ashes have been found to increase the sulphate resistance of concrete. However, this may or may not be so with ASTM Class C fly ashes. Dikeou (124) reported the results of sulphate resistance studies on a total of 30 concrete mixtures made from three portland cements, 3 portland-fly ash cements, and 12 ASTM Class F fly ashes. From this work it was concluded that all of the fly ashes tested greatly improved the sulphate resistance. The relative order of improvement found in his investigation is shown in Fig. 7.2. Accelerated wetting and drying tends to promote microcracks, therefore the data from this test may not be reliable.

In 1976, Dustan (125) reported the results of experiments on 13 concrete mixtures using fly ashes from lignite and subbituminous coals. On the basis of his work, he concluded that lignite and subbituminous fly ash concrete generally exhibited reduced resistance to sulphate attack.

In 1980, Dunstan (126) published his results of a five-year study on the sulphate attack on fly ash concretes. This report includes a theoretical analysis of the sulphate attack and its causes. The basic postulate of Dunstan's thesis is that CaO and Fe_2O_3 in fly ash are main contributors to its resistance to sulphate attack or lack of it. According to the author, as the calcium oxide content of ash increases above 5%, or as the ferric oxide content of the ash decreases, sulphate resistance is reduced. To use this observation as a means for selecting potentially sulphate-resistant fly ashes, Dunstan proposed a resistance factor (R) calculated as follows:

$$R = (C-5)/F$$

where:

C = per cent CaO
F = per cent Fe_2O_3

More recent research (127,128) has shown that R value is not reliable for predicting the sulphate resistance. The sulphate

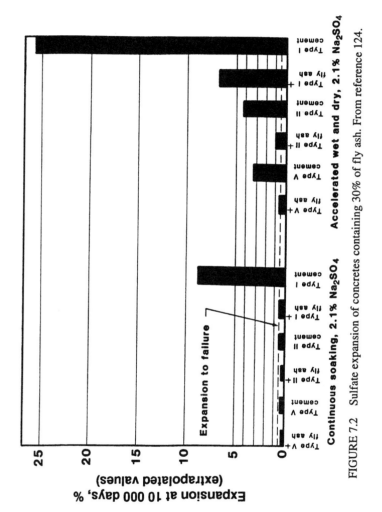

FIGURE 7.2   Sulfate expansion of concretes containing 30% of fly ash. From reference 124.

resistance depends on the amount of reactive alumina and the presence of expansive phases in the fly ash, and not on the CaO and Fe_2O_3 contents of the ash as indicated by the R factor.

ACI 232 Committee report recommends that fly ashes with less than 15 per cent CaO content will improve the sulphate resistance of concrete, and fly ashes with more than 15% CaO should be investigated for sulphate expansion using ASTM C 1012 test method.

Silica Fume

The data on the sulphate resistance of silica fume concrete exposed to 10% sodium sulphate solution were first reported by Bernhardt (129) in Norway. He concluded that the sulphate resistance of concrete improved when 10 to 15% of the portland cement was replaced by silica fume, but, he also noted that the test period was not long enough to draw any firm conclusions.

In a study by Fiskaa (130), the long-term performance of concrete specimens exposed to the ground water (containing up to 4 g/L of SO_3 with pH varying from 7 to 2.5) in Oslo's alum shale region was monitored by measuring volume changes in the concrete. The concrete specimens, most of them with a W/C of 0.50, were made incorporating various cements and additives. Results of this 20-year old study indicated that the concrete with silica fume at 15% cement replacement and W/(C + SF) of 0.63 had as much resistance to sulphate attack as a concrete made with a sulphate resisting cement at a W/C of 0.50 (Fig. 7.3). The high sulphate resistance of silica-fume concretes are attributed to the following factors:

(a) the refined pore structure, resulting in reduced transport of harmful ions;

(b) low quantity of calcium hydroxide in the hydrated paste and,

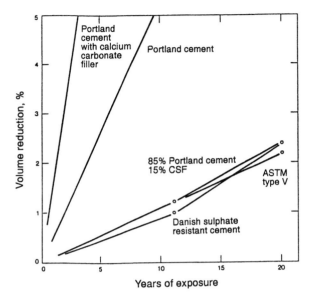

FIGURE 7.3   Volume reduction of 100 × 100 × 400-mm concrete prisms stored for 20 years in acidic, sulphate-rich water in Oslo alum-shale region. From reference 130.

(c)   increase in aluminum incorporated in the hydration products which reduces the amount of alumina available for ettringite formation.

In an investigation by Yamato, Soeda and Emoto (131), the chemical resistance of silica fume concretes in solutions of 2% hydrochloric acid, and 10% sodium sulphate was studied. The non air-entrained concretes, made incorporating 10, 20, and 30% silica fume and a W/(C + SF) of 0.55 were water-cured for 91 days before immersion in the solutions. The weight loss and the reduction in dynamic modulus of elasticity were monitored for about one year. Fig. 7.4 and 7.5 show the weight loss of concretes in these solutions. The authors concluded that the

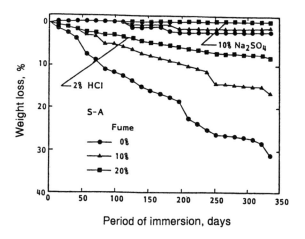

FIGURE 7.4   Effect of SF content on weight loss due to 10 percent
             Na_2SO_4 and 2 percent HCl solutions in concrete.
             From reference 131.

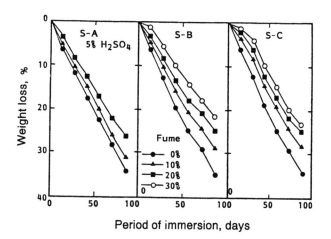

FIGURE 7.5   Effect of SF content on weight loss due to 5 percent
             H_2SO_4 solution in concrete. From reference 131.

incorporation of silica fume in concrete resulted in significant increases in chemical resistance to the above aggressive solutions. They attributed this to the reduction in permeability and the amount of potentially soluble lime.

Durning and Hicks (132) measured the resistance of silica-fume concrete to aggressive chemicals. The air-entrained concretes were made incorporating 7.5, 15, and 30% silica fume, with W/(C + SF) of 0.26 and 0.36. The specimens, 75 × 150-mm in size, were moist cured for 28 days before exposure to 1 and 5% sulphuric acid, 5% formic acid, 5% acetic acid, and 5% phosphoric acid. A mixed sodium and magnesium sulphate solution (0.175 mole/litre of each) was used to evaluate resistance to sulphate attack. Table 7.3 shows the number of cycles of wetting and drying to 25% mass loss. It was found that as the dosage rate of silica fume was increased, the number of cycles to failure increased in all solutions. The authors noted that a 30% silica fume dosage would virtually eliminate all calcium hydroxide in the cement paste. This would lead to the degradation of concrete resulting from the deterioration of the aggregate and of the calcium silicate hydrate. In the specimens exposed to the mixed sulphate solutions, no appreciable degradation was observed.

Mehta (133) compared the chemical resistance of low water-to-cement ratio (0.33 to 0.35) concretes exposed to the solutions of 1% hydrochloric acid, 1% sulphuric acid, 1% lactic acid, 5% acetic acid, 5% ammonium sulphate, and 5% sodium sulphate. Specimens of a reference concrete, latex modified concrete, and silica fume concrete with 15% silica fume by weight of cement were used, and the criteria for failure was 25% mass loss when fully submerged in the above solutions. The investigation showed that the concrete incorporating silica fume showed better resistance to chemical attacks than the other two types of concretes. The only exception was the silica fume concrete specimens, immersed in either 1% sulfuric acid or 5% ammonium sulphate solution, which performed poorly. This was attributed to the ability of the corro-

TABLE 7.3 Resistance of Concrete Containing Silica Fume to Chemical Attack*

Sample Code	Silica Fume, %	Environment	Cycles to 25% mass loss	Percent Loss at 25 cycles
1SA	0	1% sulfuric acid	13	—
2SA	7.5	1% sulfuric acid	19	—
3SA	15	1% sulfuric acid	22	—
4SA	30	1% sulfuric acid	32	22
1AA	0	5% acetic acid	8	—
2AA	7.5	5% acetic acid	17	—
3AA	15	5% acetic acid	32	22
4AA	30	5% acetic acid	>60	8
1FA	0	5% formic acid	17	—
2FA	7.5	5% formic acid	31	—
3FA	15	5% formic acid	>50	12+
4FA	30	5% formic acid	>50	8+
1SA5	0	5% sulfuric acid	4	—
2SA5	12.5	5% sulfuric acid	5	—
3SA5	25	5% sulfuric acid	5.3	—
1PA	0	5% phosphoric acid	12	—
2PA	12.5	5% phosphoric acid	13.6	—
3PA	25	5% phosphoric acid	17.3	—

+30 cycles
*From reference 132

sive ions to decompose the calcium silicate hydrate in the hydrated cement paste. However, in another investigation, Popovics et al. (134) observed that silica fume inhibited ammonium sulphate corrosion. Here, mortar specimens made with portland cement, blended cement with 20% slag, blended cements with 15% natural pozzolans, and portland cement with 15% silica fume addition were exposed to 10% ammonium sulphate solution following 28 days of water curing. The authors concluded that sulphate corrosion was predominant in this solution, and silica fume was effective in reducing this type of corrosion as well as acid corrosion. This was found to be true in silica fume mortars with different porosities.

Carlsson et al. (135) reported an investigation in which 100-mm concrete cubes were half immersed in saturated, calcium nitrate and ammonium nitrate solutions. The concretes were made both with and without a silica fume admixture which contained 90% silica fume and the rest superplasticizer and portland cement. The results showed that after 108 weeks of exposure to ammonium nitrate solution, the specimens incorporating the silica-fume admixture had a weight loss of less than 1%, whereas the specimens without the admixture had a weight loss of 15.1%. When comparing the compressive strengths, it was found that specimens made with the admixture had lost less than 20% of the original strength, while the specimens without the admixture had lost more than 70% strength.

Slag

In general, blended cements containing more than 65% of slag are known to be sulphate resistant irrespective of the chemical composition of portland cement and slag. According to Ludwig (139), the cements exhibiting resistance to sulphate attack are:

- portland cement with a C_3A content ≤ 3 wt%;
- slag cement with $\geq 70\%$ slag content; and
- non-standard cements such as high-alumina and super-sulphated cements.

In Canada, two separate studies (136,137) have shown that mortar and concretes made with blended cements composed of 50% pelletized slag and 50% CSA Type 10 (ASTM Type I) normal portland cement were equivalent in sulphate resistance to concrete made with CSA Type 50 (ASTM Type V) portland cement. The results of the studies provided the technical basis for using pelletized slag cement as a mineral admixture for sulphate resistance concrete in accordance with Canadian Standards Association Standard CAN3–423. Hogan and Meusel (57) carried out a study with ASTM Type II cements that showed high resistance to sulphate attack when the granulated slag proportion exceeded 50% of the total cementitious material.

Results of recent studies carried out by Frearson (138) confirm the inferior resistance to sulphate attack of ordinary portland cements and of blends of both ordinary and sulphate-resisting portland cement containing lower levels of granulated slag replacement. Sulphate resistance increased with increasing granulated slag content, and a mortar with 70% slag content was found to have a resistance superior to those containing sulphate-resisting portland cements alone. Also, the influence of slag content on sulphate resistance was found to be more significant than the water-to-cement ratio in the range examined.

Bakker (140) found that slag concretes with a high slag content display an increased resistance to sulphates because of the decreased permeability of the concrete to different ions and water, as shown by the diffusion coefficient values.

Results of the study carried out by Hogan and Meusel (57) on one cement and on three levels of cement replacement by slag indicated that, as the slag replacement of portland cement is increased, the resistance to sulphates improved (Fig. 7.6 and 7.7). According to the data developed, the mixtures of Type II cement and granulated slag at 40 to 60% replacement resulted in blends with superior resistance to sulphate attack.

Where granulated slag is used in sufficient quantities, several changes occur that improve resistance to sulphate attack. These changes include:

- the C_3A content of the mixture is proportionally reduced depending on the percentage of slag used. However, Lea (1) reported that increased sulphate resistance not only

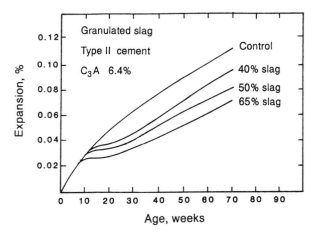

FIGURE 7.6   Sulphate resistance of mortar bars: Wolochow test methods (LTS 21 Type II). From reference 57.

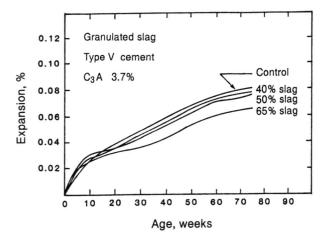

FIGURE 7.7   Sulphate resistance of mortar bars: Wolochow test method (LTS 51 Type V). From reference 57.

depends on the C_3A content of portland cement alone, but also on the Al_2O_3 content of the granulated slag. Lea further reported that sulphate resistance increased where the alumina content of the slag was less than 11%, regardless of the C_3A content of the portland cement when blends with 20–50% granulated slags were used.

- through the reduction of soluble $Ca(OH)_2$ in the formation of calcium silicate hydrates (CSH), the environment for the formation of ettringite is reduced.

- resistance to sulphate attack is greatly dependent on the permeability of the concrete or cement paste (140,141, 142). The formation of CSH in pore spaces, usually occupied by alkalis and $Ca(OH)_2$, reduces the permeability of the paste, and prevents the intrusion of aggressive sulphates.

Mehta (143) tested pastes incorporating natural pozzolans, rice-husk ash, and granulated slag. The 28-day old paste of the blended cement containing 70% blast-furnace slag showed excellent resistance to sulphate attack. There were hardly any large pores present in the hydrated paste, although the total porosity (pores >45Å) was the highest among all the cement tested. The direct relationship between sulphate resistance of a cement and the slope of its pore-size distribution plot in the range 500–45Å, probably shows that the presence of a large number of fine pores is associated with improved sulphate resistance of the material (Fig. 7.8). Although the total porosity of the cement containing 30% slag was considerably less than the cement containing 70% slag, the former was not found to be sulphate resistant. On the basis of the test results, Mehta proposed that the chemical resistance of blended portland cements results mainly from the process of pore refinement which is associated with the pozzolanic reactions involving the removal of $Ca(OH)_2$.

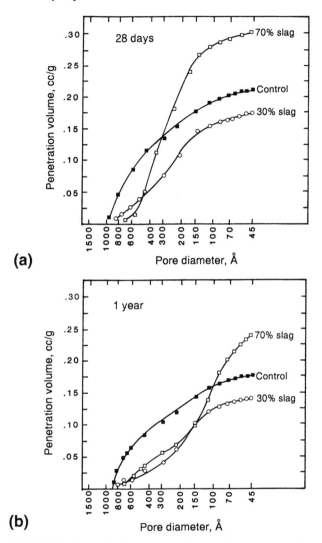

FIGURE 7.8 Pore-size distribution of hydrated cements containing 30 or 70% granulated blast-furnace slag (a) 28-days, (b) 365-days. From reference 143.

Natural Pozzolans

Use of natural pozzolans with portland cement in concrete generally increases the resistance to sulphate-bearing soil and waters. The relative improvement is greater for concrete with low cement content. The use of pozzolans with sulphate-resisting portland cements may not increase sulphate resistance, and if chemically active aluminum compounds are present in the pozzolan, a reduction in sulphate resistance of the concrete may result. In a series of tests of 20 cements and blends of Type I with Class F fly ash, Santorin earth, and silica fume, it was found that blended cements manufactured using highly siliceous natural or artificial pozzolans, slags, or silica fume performed better in sulphate environments than ordinary portland cements having the same C_3A content (144).

DURABILITY OF CONCRETE TO REPEATED CYCLES OF FREEZING AND THAWING

The freezing and thawing resistance of concrete is generally investigated using ASTM Standard C 666. The method is designated as rapid because it allows for alternatively lowering the temperature of specimens from 4.4 to $-17.8°C$ and raising it from -17.8 to $4.4°C$ in not less than 2 nor more than 4 h. The customary accepted duration of testing is 300 cycles, which can be completed in 25 to 50 days. In Procedure A of ASTM C 666, both freezing and thawing occur with the specimens surrounded by water, while in Procedure B of the test the specimens freeze in air and thaw in water; the Procedure B is less severe than Procedure A. The requirements for Procedure A are met by confining the specimen and surrounding water in a suitable container. The specimens used normally are prisms not less than 76 mm nor more than 137 mm in width and depth, and between 356 and 406 mm in length.

Extensive laboratory and field experience in Canada and the U.S.A. has shown that for satisfactory performance of concrete under repeated cycles of freezing and thawing, the cement paste should be protected by incorporating air bubbles, 10 to 100 μm in size, using an air-entraining admixture. Briefly, the most important parameters concerning the entrainment of air in concrete are the air content, bubble spacing factor and specific surface. For satisfactory freezing and thawing resistance, it is recommended that air-entrained concrete should have bubble spacing factor $(\overline{L})$ values of less than 200 μm, and specific surface (α) greater than 24 mm^2/mm^3. Usually, fresh portland cement concrete incorporating between 4 to 7% entrained air by volume will yield the above values of $\overline{L}$ and α.

In general, irrespective of the type of the mineral admixture and regardless of the percentage used, the concretes incorporating mineral admixtures have excellent durability to repeated cycles of freezing and thawing provided they have reached adequate maturity, and have proper air-void system. The exception is silica fume concrete which shows poor performance in freezing and thawing testing if the percentage of silica fume used is more than 20 per cent. A detailed discussion on the performance of concrete incorporating mineral admixtures in freezing and thawing cycling follows.

Fly Ash

In general, the observed effects of fly ash on freezing and thawing durability support the view expressed by Larson (145):

> "...Fly ash has no apparent ill effects on the air voids in hardened concrete. When a proper volume of air is entrained, characteristics of the void system meet generally accepted criteria."

Larson (145), discussing some of the difficulties of interpreting the findings of much of the early work on freezing and thawing resistance of fly ash concrete, made the following observation:

> "Fly ash concrete durability characteristics are influenced and obscured by all the factors operating on ordinary concrete. They are also related to variations in the fly ash itself and perhaps to the associated phenomenon of increased air-entraining agent requirement. When valid comparisons are made with equal strengths and air contents, however, there are no apparent differences in the freezing and thawing durability of fly ash and non fly ash concretes."

Another aspect of freezing and thawing testing procedure has been criticized by Brown et al. (146), who made the following comments on the freezing and thawing testing of blended cements:

- "When blended cements are tested according to ASTM C 666, the standard method for measuring the freezing and thawing durability of portland cement concretes, inferior resistance is usually observed. This is probably because test initiation after only a short curing period does not make proper allowance for the generally lower rate of strength development of blended cements.

- "Freezing and thawing studies, when initiated after longer curing periods, have indicated that blended cements, due to development of strengths equivalent or superior to those of portland cements also develop superior resistance to freezing and thawing."

These points should certainly be kept in mind when consideration is being given to reports of all aspects of the durability of fly ash concrete, not merely to its frost resistance.

The investigations at the Bureau of Reclamation on the effect of pozzolans in the freezing and thawing durability of concrete show that curing conditions play a vital role in this regard (Fig. 7.9).

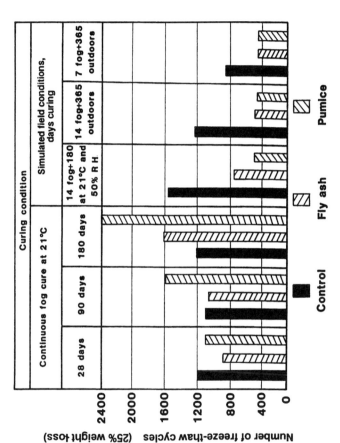

FIGURE 7.9   Effect of pozzolans on freezing and thawing durability of concrete. From reference 106.

Silica Fume

Several investigators have performed studies on the freezing and thawing resistance of silica- fume concrete. These include, among others, Sorensen (147), Gjorv (114), Malhotra and Carette (148), Malhotra (149), Yamato et al. (150), Hooton (151), Hammer and Sellevold (152), Virtanen (153), Pigeon et al. (154) and Batrakov et al. (155).

The mixture proportions, the properties of fresh concrete, and the freezing and thawing test results for one such study (149) are shown in Tables 7.4 to 7.7. The air-entrained concrete incorporated up to 30% silica fume by mass as replacement for portland cement. The W/(C + SF) was maintained at 0.40, and any loss in slump due to the use of silica fume was compensated by the use of a naphthalene-based superplasticizer. The test data revealed satisfactory performance of air-entrained concrete except for those concretes which contained 20 and 30% silica fume (Table 7.8, Figs. 7.10 and 7.11). The excessive expansions and low value of the relative dynamic modulus shown in Fig. 7.10 and 7.11 occurred in spite of the fact that the $\overline{L}$ and α values (Table 7.7) obtained were satisfactory. The author speculated that the poor performance of the concretes in question was due to the high amount of the silica fume used in concrete, resulting in a very dense cement matrix that did not permit internal movement of water (42).

In another investigation, the freezing and thawing resistance of non air-entrained and air-entrained concrete incorporating various percentages of silica fume was compared (149). Once again, the air-entrained concrete incorporating 30% silica fume failed to meet the durability criterion (Fig. 7.12), although in this case the poor performance of the concrete was attributed to the unsatisfactory $\overline{L}$ values. The above study led to the following conclusions:

"**Non air-entrained concrete:**   Non air-entrained concrete, regardless of the W/(C + SF), and irrespective of the amount

TABLE 7.4   Mix of Concrete for Freezing and Thawing Tests[*]

Mix	Replacement of cement by silica fume (%)	W/(C+SF)[A]	A/(C+SF)[B]	Quantities (kg/m^3)		AEA[C] (mL/m^3)	SP[D], % by wt of (C+SF)
				Cement	Silica Fume		
1	0	0.40	4.43	400	0	170	0.0
2	5	0.40	4.42	381	20	190	0.1
3	10	0.40	4.40	367	41	240	0.8
4	15	0.40	4.38	342	61	540	1.0
5	20	0.40	4.37	322	81	770	1.9
6	30	0.40	4.33	285	122	1090	2.7

[A]Water/(cement + silica fume) (by weight)
[B]Aggregate/(Cement + silica fume) (by weight)
[C]Air–entraining admixture
[D]Superplasticizer
*From reference 42

TABLE 7.5   Properties of Fresh Concrete in Freezing and Thawing Tests[*]

Mix	Type	W/(C+SF)[A]	Properties of Fresh Concrete			
			Temp. (°C)	Slump (mm)	Unit wt (kg/m^3)	Air content (%)
1	Control	0.40	21	75	2340	5.1
2	5% SF+SP[B]	0.40	23	65	2330	4.4
3	10% SF+SP	0.40	22	60	2370	3.8
4	15% SF+SP	0.40	22	75	2330	4.5
5	20% SF+SP	0.40	24	180[C]	2330	4.6
6	30% SF+SP	0.40	23	160[C]	2340	4.2

[A]Water/(cement + silica fume) (by weight)
[B]SP = superplasticizer
[C]Increased slumps are caused by the increased dosage of the superplasticizer
*From reference 42

TABLE 7.6 Summary of Test Results After Freezing and Thawing*

Mixture	Type	W/C +SF[A]	At the Beginning of Freezing and Thawing Cycles			At the End of Freezing and Thawing Cycles						
			Weight (kg)	Longitudinal Resonant Frequency (Hz)	Pulse Velocity (m/sec)	No. of Cycles	Weight (kg)	Longitudinal Resonant Frequency (Hz)	Pulse Velocity (m/sec)	Length Change (%)[B]	Relative Dynamic Modulus (%)[C]	Residual Flexural Strength (%)[C]
1	Control	0.40	7.402	5175	4660	600	7.289	5200	4780	0.028	101	74
2	5% SF+SP	0.40	7.335	5200	4670	500	7.286	5200	4730	0.018	100	75
3	10% SF+SP	0.40	7.350	5250	4720	500	7.329	4900	4610	0.059	87	66
4	15% SF+SP	0.40	7.327	5125	4540	525	7.289	5000	4530	0.068	95	78
5	20% SF+SP	0.40	7.285	5150	4590	425	7.251	4050	3810	0.206	62	49
6	30% SF+SP	0.40	7.224	5150	4660	300	7.205	4250	3820	0.170	68	32

[A] Water/(cement + silica fume) (by weight)
[B] Gauge length = 358 mm
[C] At the end of respective cycling, residual strength was determined in relation to reference moist-cured specimens of same age.
*From reference 42

TABLE 7.7 Air–void Characteristics of Hardened Concrete for Freezing and Thawing*

Mix-ture	Type	Air Content (%)^A	Air/paste Ratio	Specific Surface (mm^{-1})	Spacing Factor $\bar{L}$(μm)
1	Control	4.5 (5.1)	0.176	22.3	221
2	5% SF+SP	4.4 (4.4)	0.166	29.4	172
3	10% SF+SP	3.6 (3.8)	0.135	16.3	340
4	15% SF+SP	4.3 (4.5)	0.163	19.0	268
5	20% SF+SP	4.6 (4.6)	0.178	17.1	285
6	30% SF+SP	4.4 (4.2)	0.172	17.3	288

^{A}Values in parentheses refer to air content of fresh concrete
*From reference 42

of silica fume shows very low durability factors and excessive expansion when tested in accordance with ASTM C 666 (Procedure A or B). The concrete appears to show somewhat increasing distress with increasing amounts of the fume. Therefore, the use of non air-entrained condensed silica fume concrete is not recommended when it is to be subjected to repeated cycles of freezing and thawing.

Air-entrained concrete: Air-entrained concrete, regardless of the W/(C + SF) and containing up to 15% silica fume as partial replacement for cement, performs satisfactorily when tested in accordance with ASTM C 666 Procedures A and B. However, concrete incorporating 30% of the fume and a W/(C + SF) of 0.42, performs very poorly (durability factors less than 10) irrespective of the procedure used. This is probably due to the hardened concrete having high values of $\bar{L}$ or due to the high amount of condensed silica fume in concrete, resulting in a very dense cement matrix system that, in turn, might have adversely affected the movement of water. It was difficult to entrain more than 5% air in the above type of concrete and this amount of air may or may not provide satisfactory $\bar{L}$ values in hardened concrete for durability purposes.

TABLE 7.8   Mixture Proportions and Properties of Fresh and Hardened High–Strength Concrete*

Mixture No.	W/(C + SF)	A/(C + SF)	Cement (kg/m³)	Silica Fume %	Silica Fume kg/m³	Super-plasticizer (kg/m³)	Properties of Fresh Concrete Slump (mm)	Air content (%)	Density (kg/m³)	Compressive strength at 28 days (MPa)	Freezing and Thawing Test Results No. of Cycles Completed	Durability Factor
1	0.35	5.2	377	0	0	4.8	165	2.0	2460	51.5	66	6
2	0.35	5.1	342	10	38	6.2	140	1.8	2460	61.4	70	6
3	0.35	5.1	306	20	77	9.4	165	1.2	2470	68.5	70	10
7	0.30	4.3	443	0	0	7.6	215	1.9	2480	61.9	67	12
8	0.30	4.3	400	10	44	9.0	195	1.3	2475	79.3	67	3
9	0.30	4.2	356	20	89	12.5	230	1.1	2460	80.0	70	3
13	0.25	3.4	531	0	0	10.4	230	2.0	2460	65.9	89	11
14	0.25	3.4	482	10	54	11.9	230	1.3	2490	81.7	89	5
15	0.25	3.4	428	20	107	14.2	215	1.5	2465	87.1	89	8

Note: All ratios by weight; compressive strength was determine on 15 x 300–mm cylinders; freezing and thawing test used was Procedure A of ASTM C666; cement type – ASTM Type 1, coarse aggregate – crushed limestone, fine aggregate – natural sand, superplasticizer – a naphthalene–based product; freezing and thawing test results indicate complete failure of test prisms at less than 100 cycles.
*From reference 156.

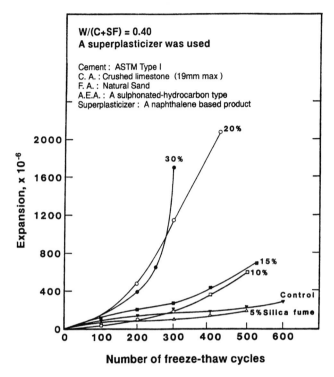

FIGURE 7.10   Expansion of test prisms after freezing and thawing
exposure (ASTM C 666 Procedure A). From refer-
ence 42.

The users are therefore asked to exercise caution when using
high percentages of condensed silica fume as replacement for
portland cement in concretes with W/(C + SF) of the order of
0.40, if these concretes are to be subjected to cycles of freezing and thawing."

Malhotra et al. (156) reported results of another investigation dealing with freezing and thawing resistance of concrete
in which non air-entrained concrete was proportioned to have a

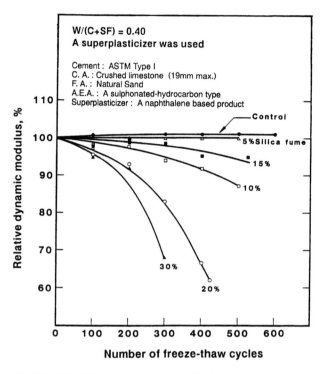

FIGURE 7.11   Relative dynamic moduli of test prisms after freezing and thawing exposure (ASTM C 666 Procedure
A). From reference 42.

W/(C + SF) ranging from 0.35 to 0.25 (156). One series of the
mixtures incorporated 10 to 20% silica fume (Table 7.8).

Large quantities of a superplasticizer were added to obtain
about 200 mm slumps. The non air-entrained concrete test
specimens, with and without the fume, when exposed to rapid
freezing and thawing tests (ASTM C 666 Procedure A) started
showing distress at less than 30 cycles, and developed major
cracks at 50 cycles. No microcracks or scaling was observed in

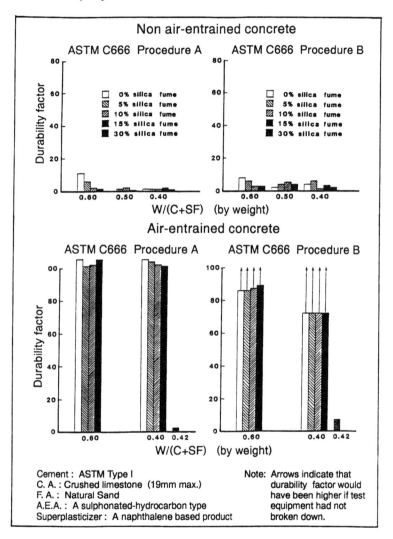

FIGURE 7.12   Durability factors for non air-entrained and air-entrained concrete. From reference 149.

any instance. The visual observations of the width of the cracks indicated that the prisms without the fume had performed marginally better. The failure of the very low W/(C + SF) concretes in the above freezing and thawing test is probably due to the availability of freezable water in the pores of high strength, very low permeability matrix having unsatisfactory strain capacity.

Data by Yamato, Emoto and Soeda (150), Hammer and Sellevold (152), and Virtanen (153) are, in general, in agreement with the above data (156). Hammer and Sellevold have tried to explain the poor performance of non air-entrained silica fume concretes (5 to 10% silica fume) with a water-to-cementitious ratio of 0.28 to 0.37 in ASTM C 666 as follows:

"There appears to be a conflict between the calorimeter data that little or no ice forms in the strongest concretes and the fact that ASTM C 666 tests showed poor performance of the same type of concrete. We suggest that thermal incompatibility between aggregate and binder may cause this, a possibility that should be investigated further."

Slag

Many studies are reported in which granulated blast-furnace slag has been used as partial replacement for portland cement in concrete subjected to cycles of freezing and thawing. Results of these studies indicate that when mortar or concrete made with granulated slag-portland cement were tested in comparison with Type I and II cement, their resistance to freezing and thawing in water (ASTM C 666, Procedure A) was essentially the same (157,158) provided the concrete was air entrained.

More recently, Hogan and Meusel (57) tested freezing and thawing durability of specimens of air-entrained concrete with 50% slag replacing portland cement, and of a control concrete

TABLE 7.9   Resistance to Repeated Cycles of Freezing and Thawing[*]

	Control	50% Slag and 50% cement
Lot	614	614
W/C	0.56	0.56
Air content, % (pressure)	5.5	6.0
Slump, mm[**]	82.6	82.6
Content, kg/m^3	149	149
Weight, kg/m^3	268	269
Change in property	2336	2326
freeze–thaw		
at 301 cycles[**]		
Expansion, %	+0.010	+0.026
Weight, %	–3.25	–2.42
Durability factor	98	89 relative durability factor = 91

[*]From reference 57
[**]1 in = 25.4 mm, 1 lb/yd3 = 0.59 kg/m3, and 1 lb/ft3 = 16 kg/m3

containing only portland cement. The specimens were stored for 14 days in a moist-curing room, and were subsequently placed in a freezing and thawing chamber. Tests were carried out using ASTM C 666, Procedure A. Although a measurable difference was found in the durability factors of the two concretes after 301 cycles of freezing and thawing, both were rated as frost-resistant, disregarding the difference in weight loss and negligible expansion (Table 7.9).

Malhotra (159) reported results of tests performed in an automatic unit capable of performing eight freezing and thawing cycles per day (ASTM C 666, Procedure B). The percentage of the slag used as replacement for normal portland cement varied from 25 to 65% by mass of cement. Initial measurements were taken at 14 days. Subsequently, two specimens were placed in the freezing and thawing cabinet, leaving two companion specimens in the moist-curing room for reference purposes. The specimens were examined visually after every 50 freezing and thawing cycling intervals. After about every 100 cycles, the specimens were measured for length change, weighed, and tested by resonant frequency and by the ultrasonic-pulse velocity method. The test was terminated after 700 cycles of freezing and thawing. Durability of the exposed concrete prisms was determined from weight, length, resonant frequency, and pulse velocity of the test prisms before and after the freezing and thawing cycling, and relative durability factors (ASTM C 666) were calculated. The test results (Table 7.10) indicate that regardless of the W/(C + S) and whether the concretes were air-entrained or air-entrained and superplasticized, these specimens performed excellently in the above tests with relative durability factors greater than 91%. Some deviation from this behaviour was justified by the fact that test prisms with slag addition were subjected to the cycling at equal ages rather than at equal compressive strengths with respect to the control specimens.

Virtanen (160) confirmed the results of earlier investigations regarding the effect of mineral admixtures on the freezing and thawing resistance of concrete. The results of his study showed that:

- in freezing and thawing tests, non air-entrained concretes incorporating slag, fly ash, or silica fume showed poorer results than portland cement concrete whereas air-entrained concrete made with slag, fly ash, or silica fume showed better frost-resistance than the corresponding portland cement concretes.

Also, the blast-furnace slag and silica-fume concretes showed higher resistance to freezing and thawing than the corresponding fly ash or portland cement concretes. It was reported that the air content of concrete had the greatest influence on the resistance of concrete to freezing and thawing. When the strength and air content were kept constant, the addition of the slag slightly improved the resistance of concrete to freezing and thawing.

TABLE 7.10   Summary of Freezing and Thawing Test Results for Concrete Incorporating Blast-Furnace Slag

Mixture series	W/C+S‡	Type of mixture	Summary of freeze–thaw test results — At zero cycles				Summary of freeze–thaw test results — At completion of 700 cycles					
			Weight, kg	Length,** mm	Longitudinal resonant frequency, Hz	Pulse velocity, m/s	Weight, kg	Length, mm	Longitudinal resonant frequency, Hz	Pulse velocity, m/s	Durability factor, %	Relative durability factor, %
B	0.38	Control + AEA	8.703	2.89	5150	4717	8.693	2.90	5200	4747	102	100
		Control + AEA + SP	8.499	2.70	5150	4684	8.486	2.72	5138	4661	99	97
		25% slag + AEA	8.697	3.00	5300	4788	8.673	3.05	5225	4788	97	95
		25% slag + AEA + SP	8.540	2.96	5125	4684	8.517	3.01	5100	4656	99	97
		65% slag + AEA	8.622	2.74	5140	4684	8.626	2.91	4950	4568	93	91
		65% slag + AEA + SP	8.302	1.59	5025	4589	8.302	1.68	4875	4531	94	92
D	0.56	Control + AEA	8.331	2.56	5000	4568	8.299	2.56	5010	4600	100	—
		Control + AEA + SP	8.443	2.76	4980	4568	8.394	2.76	4980	4504	100	100
		25% slag + AEA	8.451	2.85	5000	4573	8.416	2.88	5000	4606	100	100
		25% slag + AEA + SP	8.544	2.83	5040	4639	8.483	2.91	5050	4622	100	100
		65% slag + AEA	8.465	2.61	4950	4546	***	2.88	****	****	—	100
		65% slag + AEA + SP	8.471	2.52	4930	4563	***	2.75	****	****	70	59

†From reference 159.
*Water-to-cement + slag ratio.
**Gauge length of 345 mm should be added to this value to arrive at the exact length.
***Prisms failed at the end of 533 freezing and thawing cycles when the resonant frequency was 3840 Hz.
****Prisms failed at the end of 450 freezing and thawing cycles when the resonant frequency was 4150 Hz.

Pigeon and Regourd (161) studied the freezing and thawing resistance of cements with various amounts of granulated blast-furnace slag. The freezing and thawing resistance was estimated by the change of length, mass, and modulus of elasticity during freezing and thawing cycling. It was concluded that the observed good performance of the slag cements resulted from the dense and uniform structure of the hydrated pastes with fine pore structure.

Rice-Husk Ash

Recent investigations by Zhang and Malhotra* have shown that portland cement concrete incorporating 10% rice-husk ash as replacement for cement has excellent resistance to freezing and thawing cycling as tested in ASTM C 666, Procedure A. The durability factor after 300 cycles was 98.

FIRE RESISTANCE

With respect to exposure of concrete to sustained high temperatures, Carette et al. (162) indicated that the use of fly ash in concrete did not change the mechanical properties of the concrete in comparison to similar concrete containing only portland cement when exposed to sustained high-temperature conditions ranging from 75 to 600°C (170 to 1110°F).

Effects of high temperatures on the behaviour of silica-fume concrete are essentially similar to those observed with portland cement concrete. Following early reported cases of disintegration of silica-fume concrete at temperatures in the range of 300 to 400°C, there was a suspicion that silica fume might have contributed to a reduction in the fire resistance of concrete. However, these cases included concretes made with exceptionally low W/(C + SF), and it was suggested later that

---

*CANMET Division Report MSL 95–007 (OP&J)

this was not unusual for ultra high-strength concrete and concrete with very low permeability because of the high vapour pressure build up inside the concrete specimens on heating.

A study by Shirley, Burg and Fiorato (163) has shown that the fire resistance of silica-fume concrete of normal structural quality is in fact as good as that of normal portland cement concrete. From his review of various investigations on the fire resistance of high-strength concrete, Jahren (164) concluded that there was little or no evidence that mineral admixtures such as silica fume have an adverse effect on the fire resistance of high-strength concrete, the important factors being the physical characteristics of the system after hardening and the conditions under which it is used.

ALKALI-AGGREGATE REACTIVITY

In 1940, it was reported by Stanton (165) that a chemical reaction involving aggregates in concrete can result in abnormal expansion and cracking of concrete. This reaction known as alkali-silica reaction involves a reaction between hydroxyl (OH) ions normally associated with alkalies (Na_2O and K_2O) in the cement paste and certain reactive silica minerals present in the coarse or fine aggregates.

Certain carbonate rocks are also prone to attack by alkalis, resulting in detrimental expansion and cracking of concrete. This detrimental reaction is usually associated with argillaceous dolomitic limestone, and is designated as the expansive alkali-carbonate reaction. According to Swenson and Gillott (166), the reaction involves destruction of dolomitic crystals by the alkaline solutions, and the release of entrapped active clay minerals that absorb water directly and swell. The principal reaction products include calcite and films of Ca-Mg hydroxides and silicates.

One of the most effective methods of controlling the alkali-aggregate reactions in concrete is to use low-alkali cements.

However, this is not always possible or economical. In such situations mineral admixtures i.e., natural pozzolan, fly ash, and granulated blast-furnace slag have been used successfully to control the expansion associated with the alkali-silica reaction.

There are many hypotheses concerning the mechanisms by which the use of mineral admixtures contributes to the control of alkali-silica reactions. These may be summarized as follows:

(a) Alkali dilution caused by the partial replacement of portland cement by a mineral admixture. (Note that insoluble alkalies present in most mineral admixtures do not increase the hydroxyl ion concentration of the pore solution)

(b) Pozzolanic reaction reducing the hydroxyl ion concentration results in the formation of C–S–H with low CaO/SiO_2 ratio which has increased capacity to incorporate Na_2O and K_2O in its structure.

(c) Dense microstructure and reduced permeability with resulting lower water absorption which is necessary for the expansion of alkali-silica gel.

The amount of mineral admixture to be used for controlling the alkali-aggregate expansion would depend upon the type of reactive aggregate, the exposure conditions, the alkali-content of the cement and the reactive silica content of the aggregate, the type of mineral admixture used, and the water-to-cementitious materials ratio of the mixture. Published data indicate that the percent replacement of cement by the mineral admixture may range from 10 to 15 percent for silica fume or rice-husk ash, 20 to 30 percent for natural pozzolans, 25 to 35 percent for fly ashes and 40 to 50 per cent for blast-furnace slags. Figs. 7.13 to 7.16, which are self-explanatory, show results of investigations involving various reactive aggregates and mineral admixtures (74,57,106,167).

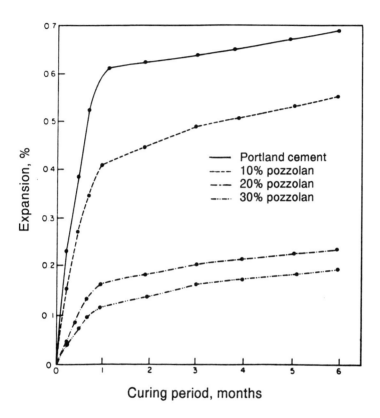

FIGURE 7.13   Control of alkali-silica expansion by Santorin earth.
From reference 74.

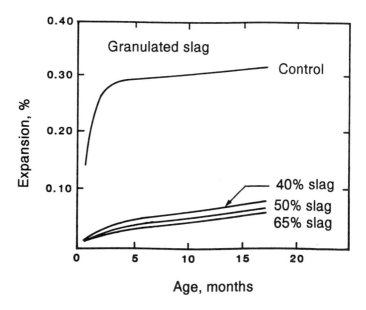

FIGURE 7.14   ASTM Test C 227 potential alkali-aggregate reactivity for various slag replacements. From reference 57.

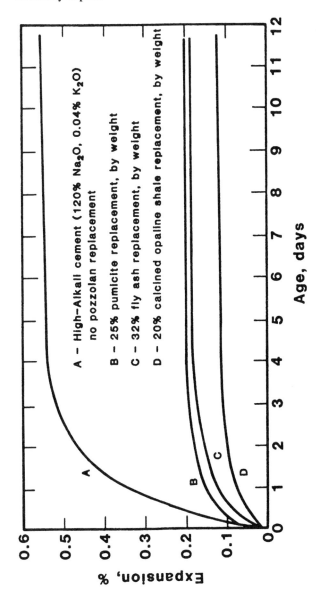

A – High–Alkali cement (120% Na₂O, 0.04% K₂O)
no pozzolan replacement

B – 25% pumicite replacement, by weight

C – 32% fly ash replacement, by weight

D – 20% calcined opaline shale replacement, by weight

FIGURE 7.15   Effect of pozzolan on reactive expansion of mortar made with alkali cement and crushed Pyrex glass. From reference 106.

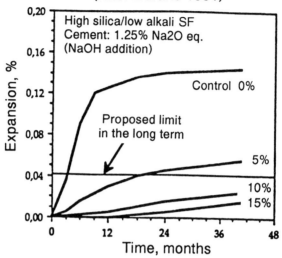

Concrete prisms with Spratt limestone
(After Durand 1991)

FIGURE 7.16   Results of expansion tests on concrete prisms (CSA
              Method) made with three reactive aggregates from
              Ottawa (Spratt Limestone), Trois-Rivières (lime-
              stone) and Sherbrooke (Chloritic schist), respective-
              ly, and various amounts of a high silica/low alkali
              condensed silica fume. For all mixtures, the alkali
              content was increased to 1.25% of the mass of ce-
              ment (Na_2O eq.) by adding NaOH to the mixture
              water. After 1 year, all samples satisfied the CSA
              limit of 0.04%. After two years, 5% SF was not suf-
              ficient for the two limestones, while the prisms
              made with the chloritic schist and containing 5%
              and 10% CSF started to expand quite rapidly. From
              reference 167.

Concrete prisms with Trois-Riv. limestone
(After Durand 1991)

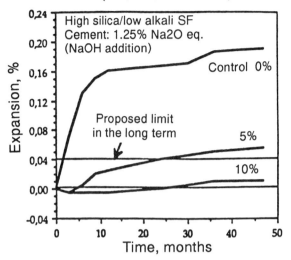

Concrete prisms with chloritic schists
(After Durand 1991)

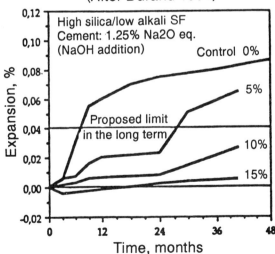

ABRASION RESISTANCE

The compressive strength, curing, and finishing of concrete, and the quality of aggregate are the major factors controlling the abrasion resistance of concrete. All other factors being equal, at comparable compressive strength, concrete with and without fly ash will exhibit essentially similar resistance to abrasion. The use of silica fume is generally associated with improvement in the compressive strength and bond strength between the cement paste and aggregate; this explains why silica fume concrete has been observed to exhibit better abrasion resistance than comparable portland cement concrete (168).

CHAPTER 8

Selection of Materials and Mixture Proportions

Each mineral admixture is unique, and its selection will depend upon the construction project requirement, availability, and cost. In principle, when one is approaching the beginning of the 21th century, no concrete should be made and placed without the incorporation of mineral admixtures. In this regard the following statement by Abdun-Nur a noted concrete technologist from the U.S.A. should be of interest (11).

"In the real world of modern concrete fly ash is as essential an ingredient of the mixture, as are cement, aggregates, water and chemical admixtures. In most concretes, I use it in larger amounts (by volume) than portland cement, and therefore it is not an admixture i.e. an addition to the mixture. Concrete without fly ash and chemical admixtures should only be found in museum showcases."

All pozzolans and granulated blast-furnace slags when used correctly in concrete increase its resistance to chemical attack, freezing and thawing cycling, and reduce expansion due to alkali-silica reaction in varying degrees. Thus when durability of concrete is of paramount concern, it should always incorporate mineral admixtures.

In general, fly ash (ASTM Class F) should be used when compressive strength at early ages is not critical, and primary aim is to reduce heat of hydration in mass concrete structures such as in dams, bridge piers, abutement walls and mat founda-

157

tions. The percentages of fly ash as a replacement for portland cement can range from 30 to 60 percent by mass of portland cement. When both an early-age strength i.e. strength at 1 and 3-day is critical, and also a reduction in the heat of hydration is required, then the approach should be to use 15 to 20% percent fly ash as replacement for cement. Alternatively, the use of high-volume fly ash concrete with low water-to-cementitious ratio (i.e. concrete incorporating a superplasticizer and about 54–58 percent fly ash as replacement for cement) may be considered (11).

ASTM Class C fly ash i.e. high-calcium fly ash should be used primarily for structural concrete where strength is the primary criterion, and high heat of hydration is not an issue. The percentage replacement levels can range from 15 to 50 percent, depending upon the strength requirements.

As mentioned in Chapter 3 of this book, unlike ASTM Class F fly ash which is pozzolanic and ASTM Class C fly ash which is both pozzolanic and cementitious, blast-furnace slag is cementitious. The ideal application for the use of slag is in structural concrete, and its use in massive structural elements to reduce heat of hydration is not recommended. The percentage replacement of portland cement by blast-furnace slag can range from 15 to 50 percent depending upon the early-age strength requirements.

The primary application of silica fume is in situations where the principal requirements are high-strength or very low permeability at early ages or both, though it has been successfully used elsewhere. For example, silica fume concrete has been successfully used for parking structures, columns in high-rise buildings to obtain strengths ranging from 100 to 115 MPa at 90 days, and in chemical plants to counteract chemical attack. The recommended percentage replacement level of portland cement by silica fume is 8 to 10 percent by weight; higher percentages of silica fume have been successfully used in specialized applications such as in repair work.

Like ASTM Class F fly ash, natural pozzolans can be used in small quantities for structural concrete, and in large quantities to reduce heat of hydration in mass concrete.

The use of mineral admixtures in concrete is not without problems. The uniformity of these materials in some instances may be questionable. The uniformity of fly ash depends mainly on the mode of operation of a power plant, with best results obtained in base load plants. The control of quality of concrete is compounded due to batching of additional and sometimes much finer ingredients; the use of silica fume is one such example. Also, the use of ultra-fine materials creates environmental problems that can only be solved by using properly designed materials handling equipment.

In the case of fly ash, silica fume, and to a lesser degree slag, air-entraining admixture demand to entrain a given volume of air in concrete is increased dramatically. There is, for example, difficulty in controlling and maintaining the entrained-air content when superplasticizers are used in concrete incorporating fly ash and silica fume. Also, at early ages concrete strength development is slower for slag/fly ash concrete except when using high-volume fly ash concrete. This is particularly troublesome in winter concreting, and this aspect has to be taken into consideration in scheduling form removal.

In general, fly ash and silica fume are darker in colour compared with portland cement. Thus, fly ash and silica fume concretes may be darker than portland cement concrete. This will have to be taken into consideration in making concrete for architectural purposes. Silica fume concrete shows increased tendency to plastic shrinkage cracking; extra precautions will therefore have to be taken to avoid rapid water evaporation from the surfaces of freshly placed silica fume concrete.

ACI 211.1–89 Standard Practice (21) outlines methods for selecting and adjusting proportions for normal-weight concrete both with and without pozzolanic and slag materials. One method is based on the estimated weight of concrete per unit volume; the other is based on calculations of the absolute volume occupied by the concrete ingredients. The trial batches are mandatory for evaluating and proportioning concrete mixtures containing mineral admixtures. By evaluating their effect on strength, water requirement, time of set, chemical admixture demand and other important durability properties, the optimum amount of total cementitious materials can be calculated. It has been reported that with the use of fly ash and granulated, blast-furnace slag, the water requirement is generally lower than that for concrete using only portland cement; on the other hand, the demand for chemical admixtures such as air-entraining and water-reducing admixtures and superplasticizers is generally higher. Furthermore, due to the differences in their specific gravities, a given weight of mineral admixture will not occupy the same volume as an equal weight of portland cement. Thus the yield of concrete would be different, and this will have to be adjusted using the actual specific gravities of the materials used.

Laboratory tests should be performed to determine the effect of the use of mineral admixtures on the setting time of concrete. The incorporation of most mineral admixtures usually slows the setting of concrete; this is further prolonged by cold weather and the use of higher percentages of the mineral admixtures. The reader is referred to ACI 211.1–89 for detailed mixture proportioning procedures. Typical mixture proportions are shown in many tables throughout this book. For proportioning high-performance concrete mixtures (65–120 MPa compressive strength) containing mineral admixtures, a reference can be made to Mehta and Aitcin (169) step-by-step procedure.

CHAPTER 9

Standard Specifications

ASTM C 618, **Standard Specification for Fly Ash and Raw or Calcined Natural Pozzolans for Use as a Mineral Admixture in Portland Cement Concrete**, covers the following three classes of mineral admixtures:

Class N: Raw or calcined natural pozzolans such as diatomaceous earth, opaline chert, tuffs, volcanic ashes (pumicite), and calcined clays or shales.

Class F: Fly ash produced from the combustion of anthracite or bituminous coal.

Class C: Fly ash produced from the combustion of lignite or sub-bituminous coal.

Originally, the ASTM Standard Specification covered only raw or calcined natural pozzolans. As new materials with similar performance characteristics appeared on the scene, the standard was amended to accommodate their use. For instance, Class F fly ash and Class C fly ash were added to ASTM C 618 in 1945 and 1975, respectively. This is why ASTM C 618 classifies the mineral admixtures by their source, and places a heavy emphasis on prescriptive chemical and physical requirements. Iron blast-furnace slag and silica fume are covered by separate standards which were issued in 1981 and 1993, respectively.

The advantages of performance-oriented specifications over prescriptive specifications will become obvious from a review of ASTM C 989, Standard Specification on Iron Blast-furnace Slag. This standard is free from the cumbersome

161

chemical and physical requirements that dominate ASTM C
618. As explained below, the prescriptive chemical and physical requirements are not of much help in predicting the performance of a mineral admixture in portland cement concrete.
Moreover, they are an obstacle to the development of a unified
standard specification covering all the mineral admixtures.
The Canadian Standard Specification CSA–A23.5 (1986) discussed below, is an example of how natural pozzolans, iron
blast-furnace slag, fly ashes, and silica fume can be included in
a single standard specification.

It is not necessary to present here a review of many world
standards on mineral admixtures as most standards have certain common features. In this chapter, highlights from the Canadian, the U.K., and the U.S. standard specifications for fly
ash, natural pozzolans, iron blast-furnace slag, and silica fume
are briefly reviewed. Chemical and physical requirements,
both that are meaningful and unnecessary from standpoint of
predicting the performance of the admixture in concrete, are
identified. It is attempted to provide a clear understanding of
the significance and limitations of the specification requirements which is essential for the development of a perfor-
mance-based specification covering all mineral admixtures.

FLY ASH

According to the Canadian, the U.K., and the U.S. Standards,
the mandatory chemical and physical requirements for fly ash
to be used as a mineral admixtures with portland cement con-
crete, are summarized in Tables 9.1 and 9.2, respectively.
Note that the U.K. Standard does not differentiate between
Class C and Class F fly ashes. The U.S. and the Canadian
Standards do make some differentiation as shown in
Table 9.1.

Free Moisture: The maximum limit for the free moisture
in fly ash ranges from 0.5 percent in the U.K. Standard to 3

TABLE 9.1 Chemical Requirements for Fly Ash for Use with Portland Cement Concrete

	U.K. B.S. 3892 (1993)	Canada CSA–A23.5		U.S. ASTM C 618	
		Type F	Type C	Class F	Class C
Free moisture, max. %	0.5	*	*	3.0	3.0
Loss on ignition, max. %	6.0	12.0	6.0	6.0+	6.0
$(SiO_2+Al_2O_3+Fe_2O_3)$, min. %	—	—	—	70	50
CaO, max. %	10	—	—	—	—
SO_3 , max. %	2.0	5.0	5.0	5.0	5.0

*The Canadian Standard does not contain any mandatory requirement for the free moisture content of fly ash. However, the user may exercise the option to require a maximum of 3.0% free moisture.
+The use of Class F fly ash containing up to 12% loss on ignition may be approved by the user if either acceptable performance records or laboratory test results are made available.

percent in the U.S. and the Canadian Standards. Moist fly ash is difficult to handle. Also, some of the anhydrous constituents may become partially hydrated and lose their reactivity in the presence of moisture.

Loss on Ignition: The loss of ignition in fly ash is generally attributable to unburnt carbonaceous matter which, due to high surface area, increases the water demand for obtaining standard consistency as well as the required dosage of water-reducing and air-entraining admixtures. The U.K. Standard, BS 3892–1993, has a maximum limit of 6 percent on loss of ignition, whereas the U.S. and the Canadian Standards for Class F fly ash permit a maximum of 6 percent and 12 percent loss on ignition, respectively. According to ASTM C 618,

TABLE 9.2   Physical Requirements for Fly Ash for Use with Portland Cement Concrete

	U.K. B.S. 3892	Canada CSA–A 23.5	U.S. ASTM C 618
Fineness 45 µm (No. 325) sieve residue max. %:	12.0	34	34
Water requirement max. of control:	95	—	105
Soundness Autoclave expansion, max. %:	—	—	0.8
Uniformity requirements: Specific gravity, max. % variation:	—	5	5
Fineness, 45 µm residue, max. % variation:	—	5	5
Strength activity index With portland cement, 7 days or 28 days min. % of control:	85*	68*	75*
With lime, 7 days at 55°C min. psi (MPa)	—	—	800(5.5)

*The strength test values between different standards are not
comparable because they are based on different mixture proportions
and curing conditions.

the use of Class F fly ash with up to 12 percent loss on ignition may be approved by the user if either acceptable performance record or laboratory test results are made available.
The 12% max. limit in the Canadian Standard is applicable to
Class F fly ash; for Class C fly ash a lower limit of 6% is applicable.

Chemical Oxides: ASTM C 618 differs from most other
world standards in that it prescribes a minimum limit on the

use of three oxides, SiO_2, Al_2O_3, and Fe_2O_3. For instance, a minimum of 70 percent limit is prescribed for Class F, and 50 percent for Class C. These limits are arbitrary, and numerous researchers have observed that they do not correlate well with the performance of the material in concrete. It is assumed that these constituents are present in a noncrystalline state (i.e. in the form of glass), and are therefore responsible for the pozzolanic activity or the formation of cementitious products by chemical reactions with calcium hydroxide at ordinary temperature. This assumption is not correct because, depending on the composition of the coal and burning conditions, some of the SiO_2, Al_2O_3, and Fe_2O_3 in fly ash would be invariably present in the form of non-reactive crystalline minerals, such as quartz, mullite, and hematite. Note that the U.K. standard contains a maximum limit of 10% on the CaO content of fly ash.

Expansion and cracking of concrete is also attributable to soluble sulfates and alkalies present in the mineral admixtures. Ettringite formation is usually the basis of sulfate-generated expansion, whereas alkali-aggregate reactions with certain reactive aggregates are the source of alkali-related expansion. The U.K. Standard has a maximum SO_3 limit of 2.0 percent, whereas the Canadian and the U.S. standard specifications for both fly ashes, Class F and C, have a maximum limit of 5 percent. As for the alkali content, only ASTM C 618 has an *optional limit* of 1.5 percent on total available alkalies (Na_2O equivalent), which is applicable when required by the purchaser for a mineral admixture to be used in concrete containing reactive aggregate. The maximum limits on SO_3 and alkalies are arbitrary because many fly ashes with higher than permissible SO_3 and alkalies seem to perform satisfactorily in concrete.

Fineness: There is a general agreement that pozzolanic properties of fly ash increase in direct proportion to the amount of fine particles present and that, in general, particles coarser than 45 μm (residue on No. 325 sieve) remain unreacted at ordinary temperature. Therefore, a restriction on the maximum allowable content of 45 μm particles seems justified as a simple approach toward quality assurance. Table 9.2 shows that the U.K. Standard restricts the particles > 45 μm in fly ash to a maximum of 12.0%, however both the Canadian and the U.S. Standards permit up to 34% material > 45 μm. Whereas the British Standard for the fineness of fly ash seems rather too restrictive, the North American standards are considered too liberal and therefore serve little purpose (except uniformity control), because almost all fly ashes produced in Canada and the United States can comply with the requirement.

Water Requirement: The U.K. and the U.S. Standards for fly ash have a maximum limit on water requirement to obtain standard consistency in portland cement-fly ash mixtures (for example, 95% of control in BS 3892). The Canadian Standard has no such requirement. This is because it covers other mineral admixtures, such as silica fume, which have excessive water requirement without the use of a plasticizing agent. It seems logical that, if an excessive water requirement can be corrected by the use of a plasticizing agent, then the decision whether or not to use the material should be left to the user and not to an overly restrictive specification.

Soundness: The Canadian and the U.S. Standards specify a maximum limit on expansion in a test for soundness using a portland cement-fly ash mixture that is subjected to autoclaving conditions. This test was initially developed for the evaluation of soundness of portland cements which show excessive expansion and cracking in the test when large amounts of crystalline MgO (periclase) or free CaO are present. Class F fly ashes normally do not contain any crystalline MgO or free CaO, hence the autoclave test for soundness serves no useful purpose. Class C fly ashes may contain significant amounts of free CaO, SO_3, and $3CaO \cdot Al_2O_3$. The presence of harmful amounts of these constituents in a Class C fly ash can be tested by a simpler test, viz. Le Chatelier's test, which is used in the U.K. Standard practice.

Strength activity index: The pozzolanic activity of a mineral admixture with portland is usually determined by an accelerated test (i.e. curing at a specified temperature which is higher than room temperature) on mortar specimens. Therefore, a direct comparison between specification requirements in various standards is not valid. The test method for the U.K. Standard is based on 30% (by mass) cement replacement with the mineral admixture. The Canadian and the U.S. Standards are based on ASTM C 311, which requires 35% cement replacement by volume. ASTM C 311 requires 1 day moist curing at normal temperature, followed by 6 or 27 days curing at 38°C. In the Canadian accelerated test, the mortar cubes are stored for 1 day at normal temperature, followed by 6 days at 65°C.

The use of accelerated curing temperatures above 50°C is not recommended by some researchers as it tends to distort the normal pozzolanic reactions. Also, the strength activity index test with lime (ASTM C 618) is not favored by many researchers as it does not truly reflect the strength potential of a mineral admixture activated with portland cement. A valid criticism against strength activity tests with portland cements is that the test results can be influenced by the composition of the portland cement used, and by a variable water-cementitious ratio which is determined from standard consistency tests.

NATURAL POZZOLANS

Raw or calcined natural pozzolans are similar to Class F fly ash in characteristics, therefore physical and chemical requirements for these materials are almost identical in most world standards. For example, Class N pozzolans (raw calcined natural pozzolans) in the ASTM C 618 are limited to a maximum of 3.0% free moisture, 10% loss on ignition, and

4% SO_3. Also, ASTM C 618 requires that the material must conform to a minimum 70% sum of the three principal oxides, namely SiO_2, Al_2O_3, and Fe_2O_3. Most other standards for raw and calcined pozzolans do not contain this chemical requirement because, as discussed above for fly ash, there is no proven relationship between the amount of these oxides and the pozzolanic activity of the material. According to Lea (1),

"These overall composition requirements do not appear in European specifications and indeed the U.S. minimum requirement comes close to rejecting some of the Roman pozzolans. It cannot be said that such limitations on chemical composition have much practical value, and they can be unnecessarily restrictive."

In the Canadian Standard, CSA–A23.5, Type N pozzolan covers raw or calcined natural pozzolans. The chemical requirements are almost identical to those for the Type F pozzolan (fly ash), namely that the loss on ignition and SO_3 are restricted to a maximum of 10% and 3%, respectively. The 3.0% limit on free moisture is optional, and there is no minimum requirement on the sum of principal oxides.

In regard to the physical requirements, the Canadian Standard has exactly the same physical requirements for the three types of pozzolans, namely Type N (raw or calcined natural pozzolan), Type F fly ash, and Type C fly ash. ASTM C 618 has also similar physical requirements for Class N pozzolans, and Class F and Class C fly ashes except that a higher water requirement for standard consistency is permitted for raw or calcined natural pozzolans (Table 9.2).

IRON BLAST-FURNACE SLAG

European standard specifications for rapidly-cooled iron blast-furnace slag suitable for use as a cementitious admixture with portland cement concrete are usually based on a minimum basicity ratio (sum of CaO + Al_2O_3 + MgO divided by SiO_2). This is because it is generally known that the cementitious property of slag is positively influenced by an increase in CaO, MgO, and Al_2O_3, whereas it is negatively influenced by an increase in SiO_2. The British Standard, BS 6699–1992, for iron blast-furnace slag for use in portland cement concrete requires that the slag must comply with a minimum basicity ratio of 1. This standard also requires a minimum glass content of 67%. Because of recent trends favoring the performance-oriented specifications over (CaO + MgO/SiO_2) prescriptive specifications, both the Canadian Standard Specification on Supplementary Cementing Materials (CSA–A23.5) and the ASTM Standard Specification for Ground Granulated Blast-furnace Slag for Use in Concrete (ASTM C 989) do not contain the basicity ratio requirement. The only chemical requirement in the ASTM Standard is a maximum SO_3 limit of 4.0% (2.5% max. sulfide sulfur). The Canadian Standard contains a maximum 5.0% limit on SO_3, whereas the British Standard has a 2.0% maximum limit on total sulfur (expressed as sulfide).

The strength contribution potential of slag is greatly influenced by its particle size. As particles > 45 μm (No. 325 sieve residue) are relatively inert, both the ASTM and CSA Standards contain 20% maximum limit for wet-sieved residue on 45 μm sieve. The U.K. Standard requires a minimum specific surface of 2750 cm^2/g. The most meaningful performance specification is based on the tests for slag activity index. The U.K. Standard test is performed with a slag-cement mixture containing 70% slag by mass, and the test mortar cubes are required to achieve a minimum of 12 and 32.5 MPa compressive strength, respectively, after 7 and 28 days of standard curing. The Slag Activity Index test in CSA–A23.5 requires 50% cement replacement with slag by volume in the test mortar mixture. After 28 days of normal curing, the test mixture must achieve at least 80% of the compressive strength compared to the control mixture without slag. ASTM C 989 permits slags to be graded into three categories based on the slag activity index with a reference portland cement meeting certain alkali and strength requirements. The test cement-slag mixture contains 50% slag by mass. The 28-day compressive strengths of normally cured test mortars must be at least 75, 95, and 115 percent of control cement mortars in order to qualify for slag grades No. 80, 100, and 120, respectively.

SILICA FUME

The Canadian Standard, CSA–A23.5, issued first in 1986 and reconfirmed without change in 1992, is one of the earliest standards in which silica fume, resulting from the production of silicon metal or ferro-silicon alloys containing at least 85% silicon, is covered as "Type U" supplementary cementing material. In 1993, ASTM issued C 1240, Standard Specification for Silica Fume for Use in Hydraulic-Cement Concrete and Mortar. Both standards have similar chemical and physical requirements. The chemical requirements include a minimum limit of 85% on SiO_2 content, and a maximum limit of 6.0% on loss on ignition. The ASTM Standard has a 3.0% maximum limit on free moisture, which is an optional requirement in the Canadian Standard. A 1.0% maximum limit on SO_3 is contained only in the Canadian Standard.

The mandatory physical requirements in both standards include a maximum 10% limit on particles > 45 μm (No. 325 sieve residue), and a minimum pozzolanic activity index with portland cement in an accelerated test. Mixture proportions for the control portland cement mortar and the test mortar containing silica fume are the same as in ASTM C 311 except that the test mortar contains 10% silica fume by mass of the total cementitious material. Both the control and test mortars are normally cured for 1 day followed by 6 days curing at 65°C in air-tight containers. A minimum of 85 Pozzolanic Activity Index (defined as the compressive strength ratio between the test mortar and the control mortar) is required for silica fume in both CSA–A23.5 and ASTM C 1260.

PROPOSAL FOR A SINGLE STANDARD FOR ALL MINERAL ADMIXTURES

From a critical examination of several world standards on mineral admixtures, Mehta (170) came to the conclusion that many requirements in the current standards are unnecessary, and therefore can be deleted in favor of only those that are meaningful for the purposes of quality assurance and for guaranteeing a minimum performance level with portland cement-concrete. It was concluded that a simple performance-oriented specification covering all mineral admixtures, similar to the Canadian Standard CSA–A23.5, can be developed by specifying limits for loss on ignition, fineness, and strength activity index. For instance, a mandatory requirements of 6.0% max. on loss on ignition and 20% max. on particles > 45 μm should be adequate for quality and uniformity assurance. A standard test for strength activity index with portland cement will be necessary to evaluate the strength-contribution potential of a mineral admixture. The minimum strength activity index requirement will be higher for highly active pozzolans such as silica fume, and lower for less active pozzolans such as natural pozzolans and fly ash, which is already reflected by the values prescribed in the Canadian Standard Specification.

References

1. Lea, F.M. The Chemistry of Cement and Concrete, Chemical Publishing Company, Inc., 1971, pp. 1–10.
2. Mehta, P.K. "Pozzolanic and cementitious by-products as mineral admixtures for concrete—a critical review"; ACI SP–79, 1983, pp. 1–46. Editor: V.M. Malhotra.
3. Caldarone, M.A., Gruber, K.A., and Burg, R.G. "High-reactivity metakaolin: a new generation mineral admixture"; Concrete International, V. 16, No. 11, 1994, pp. 37–40.
4. Ambroise, J., Maximilien, S., and Pera, J. "Properties of metakaolin blended cements"; Advanced Cement Based Materials, V. 1, No. 4, 1994, pp. 162–168.
5. "Fly ash as addition to concrete"; Dutch Center for Civil Engineering Research and Codes, Report No. 144, 1991, 99 pages.
6. Regourd, M. "Slags and slag cements"; Cement Replacement Materials, V. 3, Surrey University Press, U.K., 1986, pp. 73–99. Editor: R.N. Swamy.
7. Mehta, P.K. "Rice husk ash—a unique supplementary cementing material"; Advances in Concrete Technology, MSL Report 94–1 (R), CANMET, 1994, pp. 419–444. Editor: V.M. Malhotra.
8. Mehta, P.K. "Highly durable cement products containing siliceous ashes"; U.S. Patent No. 5, 346, 548, Sept. 13, 1994.
9. Mehta, P.K. "Natural pozzolans"; Supplementary Cementing Materials, CANMET Special Publication SP86–8E, 1987, pp. 1–33. Editor: V.M. Malhotra.
10. Mehta, P.K. "Influence of fly ash characteristics on the strength of portland cement-fly ash mixtures"; Cement and Concrete Research, V. 15, 1985, pp. 669–674.
11. Malhotra, V.M. and Ramezanianpour, A.A. "Fly ash in concrete"; MSL Report 94–45 (IR), CANMET, 1994, 307 pp.

12. McCarthy, G.J., Swamson, K.D., and Keller, L.P. "Mineralogy of western fly ash"; Cement and Concrete Research, V. 14, No. 4, 1984, pp. 471–478.

13. Hooton, R.D. "The reactivity and hydration products of blast-furnace slag"; Supplementary Cementing Materials, CANMET Special Publication SP86–8E, 1987, pp. 247–290. Editor: V.M. Malhotra.

14. Aitcin, P.C., Pinsonneault, P., and Roy, D.M. "Physical and chemical characteristics of condensed silica fume"; Bull. Amer. Ceram. Soc., V. 63, 1984, pp. 1487–1491.

15. Cook, D.J. "Rice husk ash"; Cement Replacement Materials, V. 3, Surrey University Press, U.K., 1986, pp. 171–196. Editor: R.N. Swamy.

16. Wada, S., and Igawa, T. "The influence of slag grain size distribution on the quality of portland blast-furnace slag cement"; Cement Assoc. Japan, Review of 20th Annual Meeting, 1966, pp. 91–93.

17. Helmuth, R. "Fly ash in cement and concrete"; Portland Cement Association, 1987, 203 pp.

18. Mehta, P.K. "Concrete: Structure, Properties, and Materials"; Prentice Hall, Inc., 1986, p. 202.

19. Idorn, G. "Concrete in a state of nature"; ACI, SP–141, 1991. Editor: M. Geiker.

20. U.S. Bureau of Reclamation "Physical and chemical properties of fly ash—Hungry Horse Dam"; U.S. Bureau of Reclamation Report CH–95, 1948.

21. ACI Committee 211: 1–89 "Standard practice for selecting promotions for normal, heavyweight, and mass concrete"; ACI Manual of Concrete Practice 211: 1–89, 1990.

22. Massazza, F., and Costa, U. "Aspects of the pozzolanic activity and properties of pozzolanic cements"; Il Cemento, V. 76, No. 1, 1978, pp. 3–18.

23. Davis, R.E., Carlson, R.W., Kelly, J.W., and Davis, H.E. "Properties of cements and concretes containing fly ash"; ACI Journal 33:577–612; 1937.

24. Compton, F.R., and MacInnis, C. "Field of fly ash concrete"; Ontario Hydro Research News, 18–21, Jan–Mar 1952.

25. Pasko, T.J., and Larson, T.D. "Some statistical analysis of the strength and durability of fly ash concrete"; Proceedings, ASTM Vol. 62, pp. 1054–1067, 1962.

26. Price, G.C. "Investigation of concrete materials for the South Saskatchewan River Dam"; Proceedings, ASTM Vol. 61, pp. 1155–1179, 1961.

27. Brink, R.H., and Halstead, W.J. "Studies relating to the testing of fly ash for use in concrete"; Proc. ASTM, 56:1161–1206; 1956.

28. Welsh, G.B., and Burton, J.R. "Sydney fly ash in concrete"; Commonwealth Engineer (Australia) 62–67; Jan. 1, 1958.

29. Rehsi, S.S. "Studies on Indian fly ashes and their use in structural concrete"; Proceedings, Third International Ash Utilization Symposium, Pittsburgh, March 13–14, 1973, Information Circular IC 8640; U.S. Bureau of Mines, pp. 232–245, 1973.

30. Owens, P.L. "Fly ash and its usage in concrete"; Journal, Concrete Society, England, Vol. 13, No. 7, pp. 21–26, 1979.

31. Brown, J.H. "The strength and workability of concrete with PFA substitution"; Proceedings, International Symposium on the Use of PFA in Concrete, University of Leeds, England; April 14–16, 1982, pp. 151–161, 1982. Editors: J.A. Cabrera and A.K. Cusens.

32. Carette, G.G. and Malhotra, V.M. "Characterization of Canadian fly ashes and their performance in concrete"; Division Report, MRP/MSL 84–137 (OP&J); CANMET, Energy, Mines and Resources Canada, 1984.

33. Johansen, R. Silicastøv i fabrikksbetong. Langtids-effekter"; Report STF65 F79019, FCB/SINTEF; The Norwegian Institute of Technology, Trondheim, Norway, May 1979.

34. Løland, K.E., and Hustad, T. "Report 1: Fresh concrete and methods of data analysis"; Report STF65 A81031; The Norwegian Institute of Technology, Trondheim, Norway; June 1981.

35. Aitcin, P.C., Pinsonneault, P., and Rau, G. "The use of condensed silica fume in concrete"; Proceedings of the Materials Research Society Annual Meeting; Boston, Mass., U.S.A.; pp. 316–325; Nov. 1981.

36. Sellevold, E.J., and Radjy, F.F. "Condensed silica fume (microsilica) in concrete: Water demand and strength development"; Proceedings, First International Conference on the Use of Fly Ash, Silica Fume, Slag and Other Mineral By-Products in Concrete; Montebello, Canada, July 31–August 5, 1983. ACI Special Publication SP-79, pp. 677–694; 1983. Editor: V.M. Malhotra.

37. Dagestad, G. Undersøkelse av samvirke mellom sement SP–30 og RP 38, tilsatt silika, flygeaske og tilsetningsmidlene P 40 og Lomar D"; NOTEBY Report No. 19844; Oslo, Norway; July 1983.

38. Meusel, J.W. and Rose, J.H. "Production of blast furnace slag at Sparrows Point, and the workability and strength potential of concrete incorporating the slag"; ACI SP–79; 867–890, 1979. Editor: V.M. Malhotra.

39. Nicolaidis, N. "Santorin earth—an active admixture in Greek portland cements"; Proceedings, Thirtieth International Congress of Industrial Chemistry; Athens, 1957.

40. Gebler, S., and Klieger, P. "Effect of fly ash on the air-void stability of concrete"; Proceedings, First International Conference on the Use of Fly Ash, Silica Fume, Slag and Other Mineral By-Products in Concrete, Montebello, Canada, July 31–August 5, 1983, ACI Special Publication SP–79, pp. 103–142; 1983. Editor: V.M. Malhotra.

41. Malhotra, V.M. "Strength and durability characteristics of concrete incorporating a pelletized blast-furnace slag"; ACI SP–79, 891–922, 1979. Editor: V.M. Malhotra.

42. Carette, G.G. and Malhotra, V.M. "Mechanical properties, durability and drying shrinkage of portland cement concrete incorporating silica fume"; ASTM J. Cement, Concrete and Aggregates, Vol. 5, No. 1, 1983, pp. 3–13.

43. Central Electricity Generating Board, PFA Data Book; London; 1967.

44. Copeland, B.G.T. "PFA concrete for hydraulic tunnels and shafts, Dinorwick pumped storage scheme—case history"; Proceedings, International Symposium on the Use of PFA in Concrete; University of Leeds, England, April 14–16, 1982, pp. 323–343, 1982. Editors: J.A. Cabrera and A.R. Cusens.

45. Johnson, B.D.G. "The use of fly ash in Cape Town RMC operations"; Proc, 5th Int. Conf. Alkali–Aggregate Reaction in Concrete, Cape Town, South Africa, March 30–April 3, 1981, Paper S252/33.

46. Malhotra, V.M. "Comparative evaluation of three different granulated slags in concrete"; Report MSL/87–50 (OP), CANMET, Energy, Mines and Resources Canada, Ottawa, 1986.

47. Grutzeck, M.W., Roy, D.M., and Wolfe-Confer, D. "Mechanisms of hydration of portland cement composites containing

ferrosilicon dust"; Proceedings, 4th International Conference on Cement Microscopy, Las Vegas, U.S.A., 1982, pp. 193–202.

48. Maage, M. "Modified portland cement"; Rep. no. STF65 A83063, Cement and Concrete Institute, The Norwegian Institute of Technology, Trondheim, Norway, October 1983.

49. Bilodeau, A. "Influence des fumées de silice sur le ressuage et le temps de prise du béton"; CANMET Rep. no. MRP/MSL 85–22 (TR), 1985, 11 pp.

50. Davis, R.E. and Klein, A. "Effect of the use of diatomite treated with air entraining agents upon properties of concrete"; ASTM, Special Technical Publication 99:913–108, 1950.

51. Lane, R.O. and Best, J.F. "Properties and use of fly ash in portland cement concrete"; Concrete International 4:81–92, July 1982.

52. Canadian Electrical Association Report 9017–G–804: High-volume fly ash concrete using Canadian fly ashes"; April 1994, 78 pp. (available from Canadian Electrical Association, Montreal, Quebec).

53. Mailvaganam, N.P., Bhagrath, R.S., and Shaw, K.L. "Effects of admixtures on portland cement concretes incorporating blast-furnace slag and fly ash"; Proceedings, First International Conference on The Use of Fly Ash, Silica Fume, Slag and Other Mineral By-Products in Concrete, Montebello, Canada, July 31–August 5, 1983, ACI Special Publication SP–79, pp. 519–537; 1983. Editor: V.M. Malhotra.

54. Dodson, V.H. "The effect of fly ash on the setting time of concrete—chemical or physical"; Proceedings, Symposium on Fly Ash Incorporation in Hydrated Cement Systems, Materials Research Society, Boston, pp. 166–171, 1981. Editor: Sidney Diamond.

55. Ramakrishnan, V., Coyle, W.V., Brown, J., Tlustus, A., and Venkataramanujam, P. "Performance characteristics of concretes containing fly ash"; Proceedings, Symposium on Fly Ash Incorporation in Hydrated Cement Systems, Materials Research Society; Boston, pp. 233–243, 1981. Editor: Sidney Diamond.

56. Pistilli, M.F., Wintersteen, R., and Cechner, R. "The uniformity and influence of silica fume on the properties of portland cement concrete"; ASTM J. Cement, Concrete and Aggregates, Vol. 6, No. 2, 1984, pp. 120–124.

57. Hogan, F.J. and Meusel, J.W. "The evaluation for durability and strength development of ground granulated blast-furnace slag"; ASTM Cement, Concrete, and Aggregates, 3(1):40–52; 1981.

58. Johansen, R. "Risstendens ved plastik svinn"; Report STF 65 A80016, FCB/SINTEF, The Norwegian Institute of Technology, Trondheim, Norway, March 1990.

59. Sellevold, E.J. "Review: Microsilica in concrete"; Project Report No. 08037–EJS TJJ. Norwegian Building Research Institute, Oslo, Norway, 1984.

60. Yuan, R.L., and Cook, J.E. "Study of a class C fly ash concrete"; Proceedings, First International Conference on the Use of Fly Ash, Silica Fume, Slag and Other Mineral By-Products in Concrete; Montebello, Canada, July 31–August 5, 1983, ACI Special Publication SP–79, pp. 307–319, 1983. Editor: V.M. Malhotra.

61. Raba, F. Jr., Smith, S.L., and Mearing, M. "Subbituminous fly ash utilization in concrete"; Proceedings, Symposium on Fly Ash Incorporation in Hydrated Cement Systems, Materials Research Society, Boston, pp. 296–306, 1981. Editor: Sidney Diamond.

62. Lamond, J.F. "Twenty-five years' experience using fly ash in concrete"; Proceedings, First International Conference on the Use of Fly Ash, Silica Fume, Slag, and Other Mineral By-Products in Concrete, Montebello, Canada, July 31–August 5, 1983, ACI Special Publication SP–79, pp. 47–69, 1983. Editor: V.M. Malhotra.

63. Clifton, J.R., Brown, P.W., and Frohnsdorff, G. "Reactivity of fly ash with cement"; Cement Research Progress, American Ceramic Society, Columbus, Ohio, Chapter 15, pp. 321–341, 1977.

64. Mehta, P.K. "Testing and correlation of fly ash properties with respect to pozzolanic behaviour"; Electric Power Research Institute Report CS–3314, Final Report Project 1260–26, Jan. 1984.

65. Dalziel, J.A. "The effect of different portland cements upon the pozzolanicity of pulverized-fuel ashes and the strength of blended cement mortars"; Technical Report 555, Cement and Concrete Association (England), March 1983.

66. Neville, A.M. "Properties of concrete"; (2nd ed.), John Wiley, New York, p. 382, 1973.

67. Bamforth, P.B. "In situ measurement of the effect of partial port-land cement replacement using either fly ash or ground granu-lated blast furnace slag on the performance of mass concrete"; Proc. Inst. Civ. Engrs., 69:777–800, 1980.

68. Carette, G.G., and Malhotra, V.M. "Long-term strength devel-opment of silica fume concrete"; Proceedings, 4th International Conference on Fly Ash, Slag and Silica Fume, ACI Special Pub-lication SP–132, 1992.

69. Sellevold, E.J. and Nilsen, T. "Condensed silica fume in con-crete: A world review"; Supplementary Cementing Materials for Concrete, CANMET Publication SP 86–8E, Energy, Mines and Resources Canada, Ottawa, Canada, 1987, pp. 167–229. Editor: V.M. Malhotra.

70. Maage, M. "Strength and heat development in concrete: Influ-ence of fly ash and condensed silica fume"; American Concrete Institute Special Publication SP–91, 1986, pp. 923–940.

71. Detwiler, Guy "High-strength silica fume concrete—Chicago style"; ACI Concrete International Vol. 14, No. 10, October 1992, pp. 32–36.

72. Bürg, R.G. and Ost, B.W. "Engineering properties of commer-cially available high-strength concrete"; Research Bulletin, R&D 104T, Portland Cement Association, Skokie, IL, 1992, 55 pp.

73. Malhotra, V.M., Carette, G.G. and Aitcin, P.C. "Mechanical properties of portland cement concrete incorporating blast fur-nace slag and condensed silica fume"; Proceedings, RILEM/ ACI Symposiums on Technology of Concrete When Pozzolans, Slags and Chemical Admixtures are Used, Monterrey, Mexico, pp. 395–414, 1985.

74. Mehta, P.K. "Studies on blended portland cements containing Santorin earth"; Cement and Concrete Research, 11:507–518, 1981.

75. Price, W.H. "Pozzolans—A review"; Journal, American Con-crete Institute, Proceedings 72, May 1975, pp. 225–232.

76. Munday, J.G.L., and Dhir, R.K. "Mix design for corresponding strength with pulverized fuel ash as a partial cement replace-ment"; Proc Int Conf on Materials of Construction for Develop-ing Countries; Bangkok, Aug. 1978, pp. 263–273.

77. Johansen, R. "Report 6: Long term effects"; Report STF65 A81031, FCB/SINTEF, The Norwegian Institute of Technology, Trondheim, Norway, June 1981.

78. Wolsiefer, John "Ultra high-strength field placeable concrete with silica fume admixtures"; Concrete International: Design and Construction, Vol. 6, No. 4, April 1984, pp. 25–31.

79. Luther, M., and Hansen, W. "Comparison of creep and shrinkage of high-strength silica fume concretes with fly ash concretes of similar strength"; American Concrete Institute Special Publication SP–114, 1989, pp. 573–591. Editor: V.M. Malhotra.

80. Gjorv, O.E. "Norwegian experience with condensed silica fume in concrete"; Paper presented at the CANMET/American Concrete Institute International Workshop on Silica fume in Concrete, Washington, D.C., 1991, pp. 47–64. (Available from CANMET, Natural Resources Canada, Ottawa).

81. Malhotra, V.M. and Carette, G.G. "Silica fume concrete: Properties, applications and limitations"; Concrete International: Design and Construction, Vol. 5, No. 5, 1983, pp. 40–46.

82. Malhotra, V.M. "Mechanical properties and freezing and thawing resistance of non air-entrained and air-entrained condensed silica fume concrete using ASTM Test C 666 Procedures A and B"; Div. Rep. MRP/MSL 84–153 (OP&J), CANMET, Energy, Mines and Resources Canada, Ottawa, 1984.

83. ACI Committee 226 Report: "Use of fly ash in concrete"; ACI 226.3R–87, 1987 (available from American Concrete Institute, Detroit, Michigan).

84. Bentur, A. and Goldman, A. "Curing effects, strength and physical properties of high–strength silica fume concretes"; Journal of Materials in Civil Engineering, ASCE, Vol. 1, No. 1, 1989, pp. 46–48.

85. Bentur, A., Goldman, A., Cohen, M.D. "The contributions of the transition zone to the strength of high quality silica fume concretes"; Proceedings, Materials Research Society Symposium on Bonding in Cementitious Composites, Boston, 1987, pp. 97–103.

86. Odler, I. and Zurz, A. "Structure and bond strength of cement aggregate interfaces"; Proceedings, Materials Society Symposium on Bonding in Cementitious Composites, Vol. 114, Materials Research Society, 1988, pp. 21–27. Editors: S. Mindess and S.P. Shah.

87. Chen, Z.Y. and Wang, J.G. "Effect of bond strength between aggregate and cement paste on the mechanical behaviour of concrete"; Proceedings, Material Research Society Symposium on Bonding in Cementitious Composites, Vol. 114, Materials Research Society, 1988, 41–47. Editors: S. Mindess and S.P. Shah.

88. Wu, K. and Zhou, J. "The influence of the matrix–aggregate bond on the strength and brittleness of concrete"; Proceedings, Materials Research Society Symposium on Bonding in Cementitious Composites, Vol. 114, Materials Research Society, 1988, pp. 29–34. Editors: S. Mindess and S.P. Shah.

89. Carles-Gibergues, A., Grandet, J., and Ollivier, J.P. Contact zone between cement paste and aggregate"; Proceedings, International Conference on Bond in Concrete, London, England, 1982, pp. 24–33.

90. Regourd, M. "Microstructure of high strength cement systems"; Proceedings of the Materials Research Society on Very High Strength Cement Based Materials, Boston, 1985, pp. 3–17. Editor: J.F. Young.

91. Gjorv, O.E., Monteiro, P.J.M., and Mehta, P.K. "Effect of condensed silica fume on the steel-concrete bond"; ACI Materials Journal, Vol. 87, No. 6, November–December 1990, pp. 573–580.

92. Burge, T.A. "High strength lightweight concrete with condensed silica fume"; American Concrete Institute Special Publication SP–79, 1983, pp. 731–745. Editor: V.M. Malhotra.

93. Robins, P.J. and Austin, S.A. "Bond of lightweight aggregate concrete incorporating condensed silica fume"; American Concrete Institute Special Publication SP–91, 1986, pp. 941–958. Editor: V.M. Malhotra.

94. Banthia, N. "A study of some factors affecting the fiber–matrix bond in steel fiber reinforced concrete"; Canadian Journal of Civil Engineering, Vol. 17, No. 4, 1990, pp. 610–620.

95. Horiguchi, T., Saeki, N. and Fujita, Y. "Evaluation of pullout test for estimating shear, flexural, and compressive strength of fiber reinforced silica fume concrete"; ACI Materials Journal, Vol. 85, No. 2, March–April 1988, pp. 126–132.

96. Ohama, Y., Amano, M. and Endo, M. "Properties of carbon fibres reinforced cement with silica fume"; Concrete International: Design and Construction, Vol. 7, No. 3, 1995, pp. 58–62.

97. Wainright, P.I. and Tolloczko, J.J.A. "Early and later age proper-
    ties of temperature cycled slag—OPC concretes"; ACI SP–91,
    pp. 1293–1322. Editor: V.M. Malhotra, 1986.
98. Ghosh, R.S. and Timusk, J. "Creep of fly ash concrete"; ACI
    Proceedings, Vol. 78, No. 5, Sept.–October 1981, pp. 351–357.
99. Bilodeau, A., Carette, G.G. and Malhotra, V.M. "Mechanical
    properties of non air-entrained, high strength concrete incorpo-
    rating supplementary cementing materials"; Division Report
    MSL 89–129, CANMET, Energy, Mines and Resources, Cana-
    da, 1989, 30 pp.
100. Yuan, R.L. and Cook, J.E. "Time-dependent deformation of
     high strength fly ash concrete"; Proceedings, International Sym-
     posium on the Use of PFA in Concrete, University of Leeds,
     England, April 14–16, 1982. Editors: J.A. Cabrera and
     A.R. Cusens, pp. 255–261, 1982.
101. Tachibana, D., Imai, Yamazaki, N., Kawai, T., and Inada, Y.
     "High-strength concrete incorporting several admixtures";
     American Concrete Institute Special Publication SP–121, 1990,
     pp. 309–330. Editor: W.T. Hester.
102. de Larrard, F. "Creep and shrinkage of high-strength field con-
     cretes"; ACI Special Publication SP–121, 1990, pp. 577–598.
     Editor: W.T. Hester.
103. Fulton, F.S. "The properties of portland cements containing
     milled granulated blast furnace slag"; The Portland Cement
     Institute, Johannesburg, 78 pp., 1974.
104. Personal communications with Prof. P.K. Mehta dated February
     1995.
105. Philleo, R.E. "Fly ash in mass concrete"; Proc. 1st Int. Symp. on
     Fly Ash Utilization; Pittsburg, March 14–16, 1967, Information
     Circular IC 8348, U.S. Bureau of Mines, pp. 69–79, 1967.
106. Elfert, R.J. "Bureau of Reclamation experiences with fly ash and
     other pozzolans in concrete"; Proc. 3rd Int. Ash Utilization
     Symposium, Pittsburg, March 13–14, Information Circular IC
     8640, U.S. Bureau of Mines, pp. 80–93, 1973.
107. Sivasundaram, V. "Thermal crack control of mass concrete";
     MSL Divisional Report MSL 86–93 (IR); CANMET Natural
     Resources Canada, 1986, 32 pp.
108. Sivasundaram, V., Carette, G.G. and Malhotra, V.M. "Super-
     plasticized high-volume system to reduce temperature rise in
     mass concrete"; Proceedings, Eight International Coal Ash Uti-

lization Symposium, Washington, D.C., October 1987, Paper No. 34.

109. Williams, J.T. and Owens, P.L. "The implications of selected grade of United Kingdom pulverized fuel ash on the engineering design and use in structural concrete"; Proceedings, International Symposium on the Use of PFA in Concrete, University of Leeds, England, April 14–16, 1982, pp. 301–313, 1982. Editors: J.A. Cabrera and A.R. Cusens.

110. Crow, R.D., and Dunstan, E.R. "Properties of fly ash concrete"; Proceedings, Symposium on Fly Ash Incorporation in Hydrated Cement Systems; Materials Research Society, Boston, pp. 214–225, 1981. Editor: Sidney Diamond.

111. Lessard, S., Aitcin, P.C., and Regourd, M. "Development of a low heat of hydration blended cement"; American Concrete Institute, Special Publication SP–79, Vol. 2, pp. 747–763. Editor: V.M. Malhotra.

112. Mielenz, R.C. "Mineral admixtures—History and background"; Concrete International 34–42; August 1983.

113. Davis, R.E. "Pozzolanic materials—with special reference to their use in concrete pipe"; Technical Memo, American Concrete Pipe Association, 1954.

114. Gjorv, O.E. "Durability of concrete containing condensed silica fume"; American Concrete Institute, Special Publication SP–79, pp. 695–708. Editor: V.M. Malhotra.

115. Markestad, A. "An investigation of concrete in regard to permeability problems and factors influencing the results on permeability tests"; Report STF 65 A 77027, The Norwegian Institute of Technology, Trondheim, Norway, 1977.

116. Plante, P. and Bilodeau, A. "Rapid chloride ion permeability test data on concretes incorporating supplementary cementing materials"; American Concrete Institute, Special Publication SP 114, Vol. 1, 1989, pp. 625–644. Editor: V.M. Malhotra.

117. Cohen, M.D. and Okek, J. "Silica fume in PCC: The effects of form on engineering performance"; Concrete International: Design and Construction, Vol. 11, No. 11, November 1989, pp. 43–47.

118. Berke, N.S., Dallaire, M.P. and Hicks, M.C. "Plastic mechanical corrosion and chemical resistance properties of silica fume (Micro Silica) concretes": ACI SP132, 1992, pp. 1125–1149. Editor: V.M. Malhotra.

119. Regourd, M. "Structure and behaviour of slag portland cement hydrates"; Proceedings, Seventh International Congress on the Chemistry of Cement, Paris, Vol. I, III–2/10:14; 1980.
120. Manmohan, D. and Mehta, P.K. "Influence of pozzolanic, slag and chemical admixtures on pore size distribution and permeability of hydrated cement pastes"; ASTM Cement, Concrete and Aggregate Journal, Vol. 3, No. 1, pp. 63–67, 1981.
121. Calleja, J. "Durability"; Proceedings, Seventh International Congress on the Chemistry of Cements; Paris, France, Vol. 1, VII–2 2/1:47; 1980.
122. Davis, R.E. "Review of pozzolanic materials and their use in concretes"; Symposium on Use of Pozzolanic Materials in Mortars and Concrete, ASTM, STP No. 99, 1950, pp. 3–15.
123. Ho, D.W.S., and Lewis, R.K. "Carbonation of concrete incorporating fly ash or a chemical admixture"; Proceedings, First International Conference on the Use of Fly Ash, Silica Fume, Slag and Other Mineral By-Products in Concrete, Montebello, Canada, July 31–August 5, 1983, ACI Special Publication SP 79, pp. 333–346, 1983. Editor: V.M. Malhotra.
124. Dikeou, J.T. "Fly ash increases resistance of concrete to sulphate attack"; Water Resources Tech Pub Research Report 23; U.S. Bureau of Reclamation, 1970.
125. Dunstan, E.R. "Performance of lignite and sub-bituminous fly ash in concrete—a progress report"; Report REC–ERC–76–1, U.S. Bureau of Reclamation, 1976.
126. Dunstan, E.R. "A possible method for identifying fly ashes that will improve the sulphate resistance of concretes"; ASTM Cem Concr and Aggr 2:20–30, 1980.
127. Mehta, P.K. "Effect of fly ash composition on sulphate resistance of mortars"; Proc. ACI, Vol. 83, No. 6, pp. 994–1000, 1986.
128. Tikalsky, P.J. and Carrasquillo, R.L. "Fly ash evaluation and selection for use in sulphate resistant concrete"; ACI Materials Journal, Vol. 90, No. 6, pp. 545–551, 1993.
129. Bernhardt, C.I. "SiO_2—stof som sementtilsetning"; Betongen 1 dag Vol. 2, No. 17, April 1952, pp. 29–53.
130. Fiskaa, O.M. "Betong i Alunskifer"; Publication No. 101, The Norwegian Geotechnical Institute, Oslo, Norway, 1973.
131. Yamato, T., Soeda, M., and Emoto, Y. "Chemical resistance of concrete containing condensed silica fume"; American Concrete

Institute Special Publication SP–114, Vol. 2, 1989, pp. 897–913. Editor: V.M. Malhotra.

132. Durning, T.A. and Hicks, M.C. "Using microsilica to increase concrete's resistance to aggressive chemicals"; Concrete International: Design and Construction, Vol. 13, No. 3, March 1991, pp. 42–48.

133. Mehta, P.K. "Durability of low water-to-cement ratio concretes containing latex or silica fume as admixtures"; Proceedings, RILEM–ACI Symposium on Technology of Concrete When Pozzolans, Slags, and Chemical Admixtures are Used, Monterrey, Mexico, 1985, pp. 325–340.

134. Popovics, K., Ukraincik, V., and Djurekovic, A. "Improvement of mortar and concrete durability by the use of condensed silica fume"; Durability of Building Materials, 2 (1984), Elsevier Science Publishers B.V., Amsterdam, Holland, 1984, pp. 171–186.

135. Carlsson, M., Hope, R., and Pedersen, J. "Practical benefits from use of silica fume in concrete"; Departmental Report A/S Scancem, Slemmestad, Norway, 1985.

136. Chojnocki, B. "Sulfate resistance of blended (slag) cement"; Report EM–52, Ministry of Transportation and Communications, Ontario, Canada, 1981.

137. Emery, J.J. "Sulfate reistance of Standard's slag cement"; Progress Report, Trow Ltd. Consulting Engineers, April 1982.

138. Frearson, J.P.H. "Sulfate resistance of combination of portland cement and ground granulated blast furnace slag"; Proceedings, Second International Conference on Fly Ash, Silica Fume, Slag and Natural Pozzolans in Concrete, Vol. 2:1495–1524, Madrid, Spain, 1986. Editor: V.M. Malhotra.

139. Ludwig, U. "Durability of cement mortars and concrete"; In: Durability of Building Materials and Components; ASTM Special Technical Publication 691–269–281; 1980. Editors: P.J. Sereda and G.G. Litvan.

140. Bakker, R.F.M. "Permeability of blended cement concretes"; Proceedings, First International Conference on the Use of Fly Ash, Silica Fume, Slag and Other Mineral By-Products in Concrete; Montebello, Quebec, Canada, Vol. 1:589–605, 1983. Editor: V.M. Malhotra.

141. Mehta, P.K. "Durability of concrete in marine environment: A review"; ACI SP–65, Performance of Concrete in Marine Environment, Detroit, pp. 1–15, 1990. Editor: V.M. Malhotra.

142. Roy, D.M. and Idorn, G.M. "Hydration, structure and properties of blast furnace slag cements, mortar and concretes"; ACI Journal 79–43:444–457, 1982.

143. Mehta, P.K. "Sulfate resistance of blended portland cements containing pozzolans and granulated blast-furnace slag"; Proceedings, Fifth International Symposium on Concrete Technology, Monterrey, Mexico, 35–50, 1981.

144. Patzias, T., 1987, "Evaluation of sulfate resistance of hydraulic cement mortars by the ASTM C1012 test method"; Concrete Durability: Katherine and Bryant Mather International Conference, SP 100, American Concrete Institute, Detroit, pp. 2103–2120. Editor: J.M. Scanlon.

145. Larson, T.D. "Air entrainment and durability aspects of fly ash concrete"; Proc ASTM, 64:866–886, 1964.

146. Brown, P.W., Clifton, J.R., Frohnsdorff, G., and Berger, R.L. "Limitations to fly ash use in blended cements"; Proc 4th Int Ash Utilization Symp; St. Louis, Mar. 24–25, 1976, ERDA MERC/SP–76/4; 518–529, 1976.

147. Sorensen, E.V. "Freezing and thawing resistance of condensed silica fume (Microsilica) concrete exposed to deicing salts"; American Concrete Institute Special Publication SP–79, 1983, pp. 709–718. Editor: V.M. Malhotra.

148. Malhotra, V.M. and Carette, G.G. "Silica fume: A pozzolan of new interest for use in some concretes"; Concrete Construction, May 1982, pp. 443–446.

149. Malhotra, V.M. "Mechanical properties and freezing and thawing resistance of non air-entrained and eir-entrained condensed silica fume concrete using ASTM Test C 666 Procedures A and B"; Div. Rep. MRP/MSL 84–153 (OP&J), CANMET, Energy, Mines and Resources Canada, Ottawa, 1984.

150. Yamato, Takeshi, Emoto, Yukio and Soeda, Masashi "Strength and freezing and thawing resistance of concrete incorporating condensed silica fume"; American Concrete Institute Special Publication SP–91, 1986, pp. 1095–1117. Editor: V.M. Malhotra.

151. Hooton, R.D. "Some aspects of durability with condensed silica fume in pastes, mortars, and concretes"; Paper presented at the International Workshop on Condensed Silica fume in Concrete, Montreal, 1987, 30 pp.

152. Hammer, T.A. and Sellevold, E.J. "Frost resistance of high-strength concrete"; Proceedings, American Concrete Institute Special Publication, SP–121, 1990, pp. 457–487.

153. Virtenan, J. "Mineral by-products and freeze-thaw resistance of concrete"; Publication No. 22:85, Dansk Betonforening, Copenhagen, Denmark, 1985, pp. 231–254.

154. Pigeon, M., Pleau, R., and Aitcin, P.C. "Freeze-thaw durability on concrete with and without silica fume in ASTM C666 (Procedure A Test Method) Internal Cracking versus scaling"; ASTM J. Cement, Concrete and Aggregates, Vol. 8, No. 2, Winter 1986, pp. 76–85.

155. Batrakov, V.G.., Kaprielov, S.S. and Sheinfeld, A.V. "Influence of different types of silica fume having varying silica content on the microstructure and properties of concretes"; ACI SP132, 1992, pp. 943–963. Editor: V.M. Malhotra.

156. Malhotra, V.M., Painter, K.E. and Bilodeau, A. "Freezing and thawing resistance of high-strength concrete with and without condensed silica fume"; CANMET Div. Rep. MRP/MSL 86–23 (J), CANMET, Energy, Mines and Resources Canada, Ottawa, February 1986.

157. Mather, B. "Laboratory test of portland blast furnace slag cements"; Journal of the ACI 54–13:205–232; 1957.

158. Klieger, P., and Isberner, A. "Laboratory studies of blended cement—Portland cement blast furnace slag cement"; Journal of PCA, Research and Development Laboratories 9(3):35; 1967.

159. Malhotra, V.M. "Strength and durability of characteristics of concrete incorporating a pelletized blast-furnace slag"; Proceedings, First International Conference in the Use of Fly Ash, Silica Fume, Slag and Other Mineral By-Products in Concrete; Montebello, Quebec, Canada, Vol. 2:892–921 and 923–931, 1983. Editor, V.M. Malhotra.

160. Virtanen, J. "Freeze-thaw resistance of concrete containing blast-furnace slag, fly ash or condensed silica fume"; Proceedings, First International Conference in the Use of Fly Ash, Silica Fume, Slag and Other Mineral By-Products in Concrete, Montebello, Quebec, Canada, Vol. 2:923–942, 1983. Editor, V.M. Malhotra.

161. Pigeon, M., and Regourd, M. "Freezing and thawing durability of three cements with various granulated blast furnace slag contents"; Proceedings, First International Conference on the Use

of Fly Ash, Silica Fume, Slag and Other Mineral By-Products in Concrete; Montebello, Quebec, Canada, Vol. 2:979–998; 1983. Editor, V.M. Malhotra.

162. Carette, G.G., Painter, K.E., and Malhotra, V.M. "Sustained high temperature effect on concretes made with normal portland cement, normal portland cement and slag, or normal portland cement and fly ash"; Concrete International 4:41–51, July 1982.

163. Shirley, S.T., Burg, R.G. and Fiorato, A.E. "Fire endurance of high-strength concrete slabs"; ACI Materials Journal, Vol. 85, No. 2, March–April 1988, pp. 102–108.

164. Jahren, P.A. "Fire resistance of high strength/dense concrete with particular reference to the use of condensed silica fume—A review"; American Concrete Institute Special Publication SP–114, Vol. 2, 1989, pp. 1013–1049. Editor: V.M. Malhotra.

165. Stanton, T.E. "The expansion of concrete through reaction between cement and aggregate"; Proceedings, American Society of Civil Engineers, Vol. 66, 1940, pp. 1781–1811.

166. Swenson, E.G. and Gillott, J.E. "Alkali-carbonate rock reaction"; Highway Research Record, No. 45, Highway Research Board, Washington, D.C., 1964, 21 pp.

167. Durand, B. "Preventive measures against alkali-aggregate reactions"; Course Manual, Petrography and Alkali-Aggregate Reactivity; CANMET, Energy, Mines and Resources, Canada, March 1991, pp. 399–489.

168. Laplante, P., Aitcin, P.C. and Vézina, D. "Abrasion resistance of concrete"; ASCE Journal of Materials in Civil Engineering, February, Vol. 3, No. 1, 1991, pp. 19–28.

169. Mehta, P.K. and Aitcin, P.C. "Principles underlying the production of high-performance concrete", ASTM, Cement, Concrete, and Aggregates. Vol. 12, No. 2, 1990, pp. 70–78.

170. Mehta, P.K., "Standard specifications for mineral admixtures—an overview", ACI, SP–91, Editor: V.M. Malhetra, 1986, pp. 637–658.

INDEX